T0254560

SpringerBriefs in Molecular Science

Biometals

More information about this series at http://www.springer.com/series/10046

Marc Solioz

Copper and Bacteria

Evolution, Homeostasis and Toxicity

 Springer

Marc Solioz
Department Clinical Research
University of Bern
Bern, Switzerland

ISSN 2191-5407   ISSN 2191-5415   (electronic)
SpringerBriefs in Molecular Science
ISSN 2212-9901   ISSN 2542-467X   (electronic)
SpringerBriefs in Biometals
ISBN 978-3-319-94438-8   ISBN 978-3-319-94439-5   (eBook)
https://doi.org/10.1007/978-3-319-94439-5

Library of Congress Control Number: 2018945876

Printed on acid-free paper

This Springer imprint is published by the registered company Springer Nature Switzerland AG
The registered company address is: Gewerbestrasse 11, 6330 Cham, Switzerland

Preface

Of the three domains of life, bacteria, archaea, and eukarya, the two prokaryotic domains represent the bulk of the earth's biomass. Prokaryotes live in all possible niches on earth and also colonize all multicellular organisms. The techniques of metagenomics have recently allowed to define microbiomes, the collection of prokaryotic species of a microbiota (a specific niche, such as human skin or the gut). Recent work revealed that human microbiomes are affected by lifestyle, diet, and disease. We are only at the beginning of understanding the intricate human–bacteria interaction and its role in human well-being. Such an understanding requires the investigation of individual microbes to understand their metabolism and their interaction with the surrounding world.

An important facet of all bacterial life is their ability to deal with the toxic, yet essential trace element copper. The study of copper homeostasis in different prokaryotic species over nearly four decades provides an in-depth knowledge of the process. In discussing prokaryotic copper homeostasis, it is important to know that the bacterial world is divided into two groups: Gram-negative and Gram-positive organisms (the latter including all archaea). The distinction is based on a simple staining procedure in which some organisms stain, while others do not. It was devised by the Danish physician Hans Christian Gram in 1884 and has remained a valid criterion to this day (Fig. 1) [1]. Gram-positive organisms only possess a single-cell membrane which is surrounded by a cell wall, and it is the cell wall which absorbs the Gram stain. In Gram-negative bacteria, the cell wall is surrounded by a second membrane, preventing the stain to reach the cell wall. Photosynthetic bacteria can occur to either group. In Gram-negative organisms, the two membranes separated by the periplasmic space call for special transport systems for solutes and biomolecules. For this reason, many membrane- and transport-related systems, including copper homeostasis, differ significantly between Gram-positive and Gram-negative organisms, but they also share some fundamental components.

The investigation of copper homeostasis in bacteria has brought to light proteins with surprising new functions, like copper-pumping ATPases or copper chaperones. The evolutionary conservation of some of these proteins from bacteria to humans

(a)

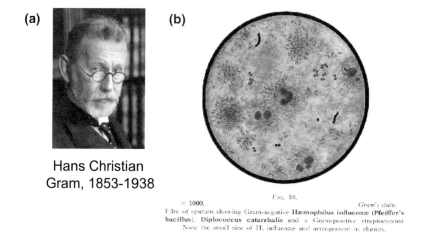

Hans Christian
Gram, 1853-1938

(b)

× 1000.   *Gram's stain.*
Film of sputum showing Gram-negative **Hæmophilus influenzæ (Pfeiffer's bacillus)**, Diplococcus catarrhalis and a Gram-positive streptococcus. Note the small size of H. influenzæ and arrangement in clumps.

(c)   E.g. *Lactococcus lactis, Enterococcus hirae*

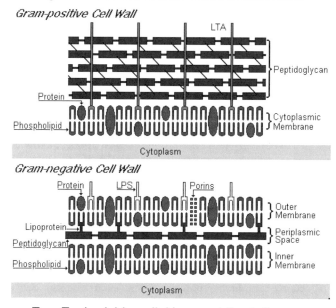

E.g. *Escherichia coli, Haemophilus influenzae*

Fig. 1 a Hans Christian Gram. **b** Gram staining of a sputum, showing swarms of Gram-negative *Haemophilus influenza*, pairs of Gram-negative *Diplococcus catarrhalis*, and two chains of Gram-positive streptococci. The few large cells are probably yeast. **c** Schematic drawing of the cell membrane-wall structures of Gram-positive and Gram-negative bacteria. Note the thick peptidoglycan cell wall of Gram-positive bacteria

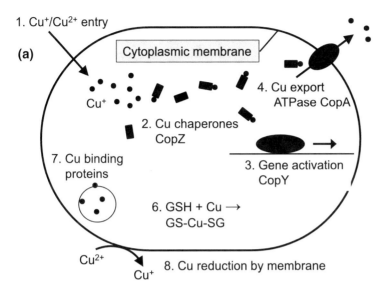

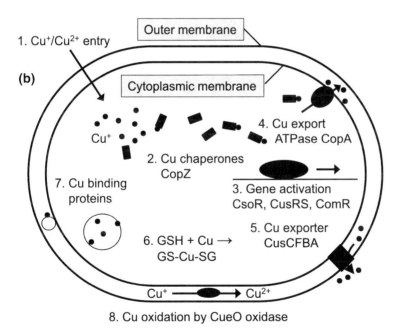

Fig. 2 Key elements of bacterial copper homeostasis. **a** Gram-positive bacterium. **b** Gram-negative bacterium. The main elements of copper homeostasis are: 1, copper entry into bacteria; 2, copper chaperones sequester cytoplasmic copper for detoxification and routing to places of export or regulations; 3, several genes are induced in response to elevated cytoplasmic copper; 4, copper is pumped across the cytoplasmic membrane by copper ATPase, powered by ATP; 5, only in Gram-positive bacteria, a CusCFBA transporter pumps copper across the outer membrane; 6, glutathione (GSH) can bind copper for detoxification; 7, copper-binding proteins buffer excess cytoplasmic copper. All steps are discussed in detail in Chaps. 3 and 4

led to a new understanding of the development of copper resistance, homeostasis, and the use of copper by cells as a modern bioelement. These evolutionary aspects will be discussed. Many examples of gene regulation by copper were characterized in detail, along with the respective genes and operons. Also, copper toxicity was revisited in recent years and new concepts have emerged. Other topics covered in this book include copper reduction by bacteria and the role of chalkophores in copper acquisition by methanotrophs, and the mechanism of copper loading of cuproenzymes. For the metallation of enzymes, I have formulated a new, universal conceptuality that first needs to be tested.

Some proteins involved in copper homeostasis are conserved from bacteria to humans, and it comes as no surprise that there has been extensive interaction, including common meetings, between researchers working on prokaryotic and eukaryotic systems. The focus of this book is laid on fundamental concepts, rather than on summarizing all the available data and historic developments (*Cupriavidus metallidurans* was the topic of a separate monograph of this series). Consequently, eukaryotic work is mentioned when it illustrates fundamental concepts particularly well or when mechanisms and proteins have been better characterized in a eukaryotic system. The copper field has greatly profited from this cross-fertilization and has made copper the trace element which is probably best understood today in terms of homeostasis and toxicity.

For a general overview and to help organize the readers' mind, Figs. 1 and 2 give simplified overviews of copper homeostasis in Gram-positive and Gram-negative bacteria. It is apparent that the two processes are of (i) limited complexity and (ii) very similar in the two bacterial worlds. The major difference is that Gram-positive bacteria require an additional copper transporter, CusCFBA, to transport copper across the outer membrane.

Keywords Copper homeostasis · Antimicrobial copper · Copper ATPases · Gene regulation by copper · Copper chaperones · Copper toxicity

Reference

Gram HC (1884) Über die isolierte Färbung der Schizomyceten in Schnitt- und Trockenpräparaten. Fortschr Medizin 2:185–189

Contents

About the Author

A native of Zurich, Marc Solioz studied chemistry and engineering in Zurich. He then moved to the USA, where he obtained his Ph.D. in Biochemistry at St. Louis University in 1975. The Ph.D. work focused on genetic transfer in photosynthetic bacteria. During postdoctoral studies at the Biocenter in Basel, he investigated the genetics and biogenesis of mitochondria. This was followed by seven years as an assistant professor at the ETH in Zurich, where he investigated proton transport by cytochrome oxidases and potassium transport by a bacterial K-ATPases.

In 1989, he became professor of biochemistry at the University of Bern. With the discovery of the first copper-pumping ATPase in 1992, his focus shifted to copper and he investigated copper homeostatic mechanisms of Gram-positive bacteria. This also brought about advisory activity on government and industry panels in relation to copper toxicity and human exposure to copper. In 2009, he started to investigate the mechanism whereby metallic copper surfaces kill bacteria.

From 2004 to 2010, he was the chief organizer of the biannual International Copper Meeting. The International Copper Meetings were inaugurated in 1997 by Arturo Leone, University of Salerno, Italy, and Julian Mercer, Murdoch Institute for Research, Melbourne, Australia, and continue to this day. These meetings have become a focal point of copper research and keep drawing copper researchers from all over the world to Italy.

Facing mandatory retirement in 2014, he moved to Tomsk State University, Siberia, where he built up a new laboratory to investigate heavy metal resistance in bacteria. Siberia offered a unique setting to isolate new heavy metal-resistant organisms from deserted mines. The endeavor was funded by a Russian Government Grant for leading scientists. This project terminated in 2016 due to funding restrictions.

Currently, he is active as a science writer and as an independent advisor to the industry. He also runs a small company which produces the antimicrobial CopperPen®.

Abbreviations

[4Fe-4S]	Iron–sulfur clusters
COX	Cytochrome c oxidase
CRP	cAMP response protein
EXAFS	Extended X-ray absorption fine structure
GSH	Glutathione (reduced)
GSSG	Glutathione (oxidized dimer)
H_2O_2	Hydrogen peroxide
HMA	Heavy metal-associated (domain)
HSAB	Pearson soft–hard acid–base concept
MBD	Metal-binding domain
Mbt	Methanobactin
MCO	Multicopper oxidase
MFS	Major facilitator superfamily
MMO	Methane monooxygenase
NAD^+	Oxidized nicotinamide adenine dinucleotide
NADH	Reduced nicotinamide adenine dinucleotide
NMR	Nuclear magnetic resonance
NO	Nitrous oxide
Pi	Inorganic phosphate
PKK1	Polyphosphate kinase
pMMO	Particulate (membrane-bound methane monooxygenase)
PPX	Exopolyphosphatase
PSII	Photosystem II
RND	Resistance–nodulation–cell division
ROS	Reactive oxygen species
SOD	Superoxide dismutase
XANES	X-ray absorption near-edge structure
Ybt	Yersiniabactin

Chapter 1
Copper—A Modern Bioelement

Abstract From the analysis of the evolution of copper-containing enzymes, it emerges that copper is a modern bioelement. It was not used as an enzyme cofactor before the advent of oxygen evolution. In the anoxic world, copper in the biosphere was in its reduced, Cu^+ state, which formed insoluble copper sulfide, promoted by the abundance of hydrogen sulfide in the atmosphere. Once the world became oxic, Cu^+ was oxidized to Cu^{2+}, which is readily soluble in the aqueous phase. The ensuing bioavailability of copper led to the evolution of cuproenzymes and copper-responsive regulators of gene expression. Indeed, all known copper-containing enzymes catalyze redox reactions involving oxygen in one form or another. Copper detoxification systems, on the other hand, have an earlier, independent evolutionary origin. The redox-active nature of copper of course makes it an ideal cofactor for redox enzymes, but also pose special experimental problems, which are discussed.

Keywords Evolution · Copper · Bioelement · Anoxic world
Cuproenzymes · Primordial · Copper sulfide

1.1 Bioavailability of Copper, Iron, Zinc and the Evolution of Species

In the anaerobic, primordial word, copper was present as copper sulfide, Cu_2S, in which copper is in the reduced Cu^+ state. This mineral, chalcocite, is insoluble in water, leaving essentially no bioavailable copper in the aqueous phase. Approximately 3 Gy ago, oxygen evolution by photosynthetic microorganisms began. It still took millions of years before oxygen started to accumulate in the atmosphere because the oxidation of Cu^+ and Fe^{2+} and other compounds initially consumed all of the produced oxygen. About 2.4 Gy ago, oxygen in the atmosphere began to rise and the dramatic geochemical changes demanded biochemical adaptions of the existing life forms to the new environments (Fig. 1.1) [1]. It may be no coincidence that the evolution of eukaryotic organisms, which started approximately 2.7 Gy ago, falls

© The Author(s) 2018
M. Solioz, *Copper and Bacteria*, SpringerBriefs in Biometals,
https://doi.org/10.1007/978-3-319-94439-5_1

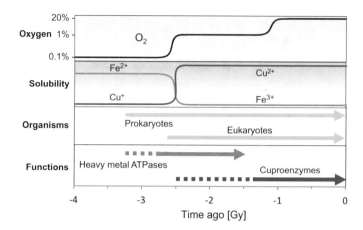

Fig. 1.1 Geochemical changes and evolution over time. The four panels schematically show the following: Oxygen, O_2 concentration in the atmosphere; Solubility, bioavailability and oxidation states of iron and copper in the oceans; Organisms, evolution of prokaryotes and eukaryotes; Functions, evolution of copper ATPases and redox cuproenzymes

into the time window of the ensuing oxygen evolution and accumulation of O_2 in the atmosphere.

Geochemically, Cu^+ exposed to the atmosphere gradually became oxidized to Cu^{2+} and thereby turned water soluble; on the other hand, iron, which had been present as water-soluble Fe^{2+}, oxidized to Fe^{3+}, which is essentially insoluble ($[Fe^{3+}] = 10^{-18}$ M at pH 7) [2]. Microorganisms growing under anaerobic conditions need iron as a redox-active enzyme cofactor for a variety of functions, such as heme formation or the reduction of ribonucleotide precursors of DNA [3]. To acquire the ever more sparse, bioavailable iron, a range of strategies evolved. Some bacteria produce siderophores (from the Greek "iron carriers"), which are low-molecular weight compounds with a very high affinity for iron [4]. Siderophores like ferrichrome, enterobactin, or citrate hydroxamate are synthesized and secreted by bacteria and fungi and scavenge iron from the environment through their very high affinity for ferric iron. Fe^{3+}-siderophore complexes are then taken up by cells and the iron is released intracellularly be reducing it to Fe^{2+}, which binds much more weekly to the siderophores. Other strategies to circumvent iron-starvation include lowering of the ambient pH, surface reduction of Fe^{3+} to more soluble Fe^{2+}, or the replacement of iron in ribonucleotide reductase by adenosylcobalmin [5]. Finally, some organisms depend on the microbial community for acquiring their iron by taking up siderophores, heme, or protein-bound iron produced by other organisms.

Another factor which may have contributed to eukaryotic evolution as the world became more oxic is the increased bioavailability of zinc. In contrast to prokaryotes, eukaryotes rely heavily on Zn^{2+}-binding proteins and they are fundamental to present-day eukaryotic cellular biology [6]. The change in global Zn bioavailability possibly prompted a burst in the evolution of proteins like Zn finger proteins, the universal transcriptional regulators of eukaryotes, accelerating the rise and diversification of Eukarya. In line with this hypothesis, zinc-binding proteins, like copper-binding proteins, evolved much later than iron-binding proteins.

As the biosphere became more oxic, organisms also had to adapt to the ensuing oxidative damage of cell components by reactive oxygen species (ROS, see Chap. 2). A key enzyme in this regard, present in most prokaryotes and eukaryotes, is superoxide dismutase (SOD). The most common form of the enzyme contains a Cu–Zn redox center, in line with the increased bioavailability of these metal ions at the time. Other enzymes which serve in the ·defense against oxidative stress are alkyl hydroperoxide reductase, thiol peroxidase, NADH peroxidase, and catalase, to name just some of the most important ones [7].

But more importantly, in the oxic world an oxidative metabolism with higher energy yield became possible and new redox enzymes for the metabolism of reduced compounds evolved. The arrival of dioxygen also created the need for a new redox active metal that could attain higher redox potentials and copper ideally suited this role. The late evolution of copper as a bioelement is also apparent from the nature of copper-containing enzymes, most of which are involved in the oxidative metabolism of highly oxidized compounds like of O_2, N_2O or NO_2^-.

Respiratory enzymes which employ copper, or copper and iron, as redox active cofactors started to appear [8]. Today, over 30 types of copper-containing proteins are known (Table 1.1) [9]. Prominent examples are lysyl oxidase involved in the crosslinking of collagen, tyrosinase required for melanin synthesis, dopamine β-hydroxylase of the catecholamine pathway, cytochrome c oxidase, the terminal electron acceptor of the respiratory chain, or superoxide dismutase, required for defense against oxidative damage, to name just a few. Most known cuproenzymes are eukaryotic, reflecting on one hand the greater genetic complexity of eukaryotes, but also their evolution in a more oxic world where copper had become readily bioavailable (Fig. 1.2) [10]. While all eukaryotic cells use copper, many bacterial species do not appear to require copper for life. In line with the concept that copper is a modern bioelement of the oxic world, these nonusers predominate among anaerobic organisms by a factor of three, but are rare (<10%) among aerobic organisms and intermediate (<50%) among facultative anaerobes [8].

It must be emphasized that the copper export ATPases, which serve to keep cytoplasmic copper below toxic levels, have an earlier and independent evolutionary origin from redox-active cuproenzymes or oxygen-binding cuproproteins [8]. Since copper can be taken up inadvertently, all microorganisms need this core detoxification system and all genomes sequenced to date contain at least one gene encoding a copper ATPase. With the increase in soluble copper, this function became even more important as a means of removing copper from the cytoplasm. The presence of copper exporters also in copper nonusers and the different pattern of occurrence of copper ATPases and cuproproteins suggests that the pathways of copper utilization and copper detoxification evolved independently of each other (see also Chap. 3) [12].

Table 1.1 Enzymes with copper redox centers[a]

Prokaryotic	Eukaryotic
Plastocyanin family (amicyanin, pseudoazurin, halocyanin, etc.)	Plastocyanin family (plantacyanin, umecyanin, mavicyanin, stellacyanin, etc.)
Azurin family (azurin, auracyanin)	
Rusticyanin	
Nitrosocyanin	
Cytochrome c oxidase subunit I	Cytochrome c oxidase subunit I
Cytochrome c oxidase subunit II	Cytochrome c oxidase subunit II
Cytochrome cbb_3 oxidase	
Cu–Zn superoxide dismutase	Cu–Zn superoxide dismutase
Copper amine oxidase	Copper amine oxidase
Nitrous oxide reductase	
Nitrate reductase	Nitrate reductase
NADH dehydrogenase 2	
Amine oxidase	Amine oxidase
Methane monooxygenase	
Multicopper oxidases (nitrite reductase, CueO, CotA, laccase, bilirubin oxidase, phenoxazinone synthase, etc.)	Multicopper oxidases (laccase, Fet3p, hephaestin, ceruloplasmin, ascorbate, oxidase, etc.)
Tyrosinase	Tyrosinase
Ribonucleotide reductase	Ribonucleotide reductase
MogA molybdenum cofactor synthesis[b]	Cnx1G molybdenum cofactor synthesis[b]
	Galactose oxidase
	Peptidylglycine R-hydroxylating monooxygenase
	Dopamine β-monooxygenase
	Galactose oxidase
	Cnx1G
	Hemocyanin

[a]Excluded from this list are proteins involved in copper transport, gene regulation, and copper chaperones. Some enzymes in the table also occur in a form that does not use copper
[b]It is not clear if copper is an absolute requirement for molybdenum cofactor synthesis (see Sect. 3.8)

1.2 The Irving-Williams Series and the Pearson Hard-Soft Acid-Base Concept

To better understand the biology of copper, it is helpful to consider some of the physico-chemical properties of copper. Irwing and Williams [13] collected data on the stability of complexes formed by the bivalent ions of the first transition series.

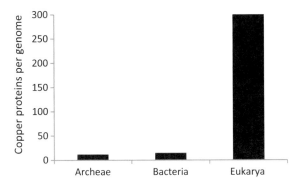

Fig. 1.2 Number of copper proteins per genome in Archaea, Bacteria, and Eukarya (The data was taken from Refs. [8, 11])

They found that the stability of such complexes almost always follows the order Mn^{2+} $<Fe^{2+}<Co^{2+}<Ni^{2+}<Cu^{2+}>Zn^{2+}$, irrespective of the nature of the coordinating ligands or the number of ligands involved. It has become fashionable in the copper field to revive the so called Irwing-Williams series to explain a protein's preference for a particular ligand over another. However, the Irwing-Williams series does not do full justice to the biology of copper. For one, Cu^+, which has properties significantly different Cu^{2+}, is not included in the series. For another, sulfhydryls were not considered as ligands by Irwing and Williams, although they are the predominant copper ligands in biology.

A better differentiation of the relevant properties of metal ions is possible based on the Pearson soft-hard acid-base concept (HSAB) [14]. Pearson's principle can be stated as follows: hard acids prefer to bond with hard bases, and soft acids prefer to bond with soft bases. The concept assigns ions, molecules and atoms with a 'hard' character as acids, and those with a 'soft' character as bases. Hard acids prefer to bond with hard bases, and soft acids prefer to bond with soft bases. Table 1.2. lists the hard-soft properties metal ions (acids) and some possible ligands (bases) which are relevant for the present discussion.

It becomes apparent from the HSAB concept and the data in Table 1.2. that in proteins, the soft acids $Au^{+/3+}$, Ag^+, Cu^+, Cd^{2+}, and Hg^{2+} preferentially bind to cysteines, methionines, or imidazole nitrogens (RNC in Table 1.2.) and that it is indeed how such metal binding sites evolved. Co^{2+}, Ni^{2+}, Fe^{2+}, Pb^{2+}, Zn^{2+}, and Cu^{2+} are intermediately soft acids, but also prefer the same ligands as the soft acids. Since Cu^+, the intracellular species of copper, is softer than the aforementioned metal ions, Cu^+ can compete with them for the same binding site, which is key to copper toxicity (see Chap. 2). The preponderance of soft metal ions to bind to thiol residues is sometimes also referred to as thiophilicity, which can be more quantitatively expressed than the hard-soft character via the solubility product, $pK_{S(MeS)}$, of the respective metal sulfides (Table 1.3) [15]. However, it must be noted that there is great variation in

Table 1.2 Hard-soft acid-base character of selected metal ions and ligands[a]

Hard acids[a]	Interm. acids[a]	Soft acids[a]	Hard bases	Interm. bases	Soft bases
Na^+, K^+	Cu^{2+}	Cu^+	ROH	Imidazole-N	RSH
Mg^{2+}	Pb^{2+}	Cd^{2+}	$RCOOPO_4^{3-}$	Aniline-NH_2	RSR
Ca^{2+}	Zn^{2+}	Hg^{2+}	RNH_2	Pyridine-N	R-phenyl
Co^{3+}	Co^{2+}	Hg^+	ROR	NO_2^-	CN^-
Fe^{3+}	Fe^{2+}	Ag^+	PO_4^{3-}, SO_4^{2-}	RN_3	H_2S
Mn^{2+}	Ni^{2+}	Au^+	NO_3^{2-}, CO_3^{2-}	N_2	H_2^-

[a]Hard-soft character according to Pearson [14]

Table 1.3 Metal sulfide solubility products

Metal ion	$pK_{S(MeS)}^a$
HgS	−53
AgS	−50
Cu_2S	−48
CuS	−37
PbS	−28
CdS	−27
SnS	−26
ZnS	−25
CoS	−22 (α) −26 (β)[b]
FeS	−19
NiS	−18
MnS	−14

[a]Data from Refs. [16, 17]
[b]Refers to the two crystal forms

the solubility constants found in the literature for some sulfides and sulfide solubility remains an approximation to thiophilicity.

1.3 Copper Speciation in the Cell and the Laboratory

The two oxidation states of copper, Cu^+ and Cu^{2+}, with a redox potential in the range of biological redox reactions, makes copper a valuable cofactor for redox enzymes. Such redox-active copper ions are deeply buried in specialized cuproenzymes and are not normally subject to exchange reactions. Metallation of these enzymes probably takes place in the periplasm and requires specialized helper proteins and enzymes (see Chaps. 3 and 4). So there may be no need for cytoplasmic copper. Indeed, free cytoplasmic copper has been estimated to be in the zeptomolar range (10^{-21} M, see Chap. 2) [18].

The cellular cytoplasm is a reducing environment. The major reducing agent in prokaryotic and eukaryotic cells is glutathione (GSH), which is present in millimolar concentrations and has a standard redox potential under cytoplasmic conditions of -0.24 V for the GSH/GSSG couple, GSSG designating the dimeric, oxidized glutathione [19, 20]. The standard redox potential, E^0, of the Cu^{2+}/Cu^+ couple is $+0.16$ V, indicating that Cu^{2+} entering the cytoplasm will immediately be reduced to Cu^+. In addition, glutathione also avidly binds copper and plays a general role in heavy metal detoxification in the cytoplasm (see Chap. 2). It can thus be assumed that all cytoplasmic copper is in the Cu^+-form and complexed by glutathione, other small molecules, or proteins. This is in line with the observation that all proteins involved in copper homeostasis, such as copper export ATPases, copper chaperones or copper regulators of gene expression specifically bind Cu^+. There has been a report of the transport of Cu^{2+} by a copper ATPase of *Archaeoglobus fulgidus*, but this remains unverified [21].

The unique properties of copper ions pose some unique challenges when working with this metal in the laboratory. First, copper ions non-specifically bind to most biomolecules, making it more difficult to single out specific effects. Secondly, work with Cu^+ requires strictly anaerobic working conditions or the use of reducing agents, which can interfere with the experiment. Specific copper chelators have become an important tool in copper research. While Cu^{2+} ions are stable in neutral, aqueous solutions exposed to the atmosphere, Cu^+ ions can only be maintained in solution at very acidic pH or in complexed form. Cu(I)-complexes which are stable even in air are formed by the chelators bicinchoninic acid, acetonitrile, or tetrathiomolybdate and these have become important research tools [22, 23]. Interestingly, tetrathiomolybdate has more recently become an alternative drug for complexing excess copper in Wilson disease patients which cannot secrete excess liver copper due to a defect in a copper-secreting ATPase [24]. Other useful copper chelators are CN^-, *o*-phenanthroline, bathophenanthroline sulfonate, or 8-hydroxyquinoline (reported formation constants for copper-phenanthroline complexes are around 21, irrespective of the ligands on the phenanthroline dipyrimidine ring system [25]). One also has to be aware that the widely used Tris-buffer is also a copper chelator [26]. Another experimental caveat is the fact that phenanthrolines complex Cu^+ (and Fe^{2+}) so strongly that they effectively raise the redox potential to a point at which any reducing equivalent in the experimental system supports the reduction of Cu^{2+} to Cu^+. This reaction can be counteracted by high concentrations (20 mM) of citrate or lactate, which preferentially bind the oxidized form of copper. Clearly, in vitro experimental work with copper requires special consideration and tools as well as great care to avoid mis-interpretations.

References

1. Crichton RR, Pierre J-L (2001) Old iron, young copper: from Mars to Venus. Biometals 14:99–112
2. Fraústo da Silva JJR, Williams RJP (1993) The biological chemistry of the elements. Oxford University Press, Oxford
3. Herrick J, Sclavi B (2007) Ribonucleotide reductase and the regulation of DNA replication: an old story and an ancient heritage. Mol Microbiol 63:22–34
4. Neilands JB (1995) Siderophores: structure and function of microbial iron transport compounds. J Biol Chem 270:26723–26726
5. Jordan A, Reichard P (1998) Ribonucleotide reductases. Annu Rev Biochem 67:71–98
6. Dupont CL, Butcher A, Valas RE et al (2010) History of biological metal utilization inferred through phylogenomic analysis of protein structures. Proc Natl Acad Sci USA 107:10567–10572
7. Baureder M, Reimann R, Hederstedt L (2012) Contribution of catalase to hydrogen peroxide resistance in *Enterococcus faecalis*. FEMS Microbiol Lett 331:160–164
8. Ridge PG, Zhang Y, Gladyshev VN (2008) Comparative genomic analyses of copper transporters and cuproproteomes reveal evolutionary dynamics of copper utilization and its link to oxygen. PLoS ONE 3:e1378
9. Kim BE, Nevitt T, Thiele DJ (2008) Mechanisms for copper acquisition, distribution and regulation. Nat Chem Biol 4:176–185
10. Dupont CL, Grass G, Rensing C (2011) Copper toxicity and the origin of bacterial resistance-new insights and applications. Metallomics 3:1109–1118
11. Zhang Y, Gladyshev VN (2010) General trends in trace element utilization revealed by comparative genomic analyses of Co, Cu, Mo, Ni and Se. J Biol Chem 285:3393–3405
12. Gladyshev VN, Zhang Y (2013) Comparative genomics analysis of the metallomes. In: Banci L (ed) Metallomics and the Cell. Springer, Heidelberg
13. Irving H, Williams RJP (1953) The stability of transition-metal complexes. J Chem Soc 1953:3192–3210
14. Pearson RG (1968) Hard and soft acid and bases, HSAB, part I. J Chem Educ 45:581–587
15. Hans M, Mathews S, Mücklich F et al (2016) Physicochemical properties of copper important for its antibacterial activity and development of a unified model. Biointerphases 11:018902-1–018902-8
16. Nies DH (2003) Efflux-mediated heavy metal resistance in prokaryotes. FEMS Microbiol Rev 27:313–339
17. Saier MH Jr, Tam R, Reizer A et al (1994) Two novel families of bacterial membrane proteins concerned with nodulation, cell division and transport. Mol Microbiol 11:841–847
18. Changela A, Chen K, Xue Y et al (2003) Molecular basis of metal-ion selectivity and zeptomolar sensitivity by CueR. Science 301:1383–1387
19. Masip L, Veeravalli K, Georgiou G (2006) The many faces of glutathione in bacteria. Antioxid Redox Signal 8:753–762
20. Fahey RC, Brown WC, Adams WB et al (1978) Occurrence of glutathione in bacteria. J Bacteriol 133:1126–1129
21. Mana-Capelli S, Mandal AK, Arguello JM (2003) *Archaeoglobus fulgidus* CopB is a thermophilic Cu^{2+}-ATPase: functional role of its histidine-rich-N-terminal metal binding domain. J Biol Chem 278:40534–40541
22. Hemmerich P, Sigwart C (1963) $Cu(CH_3CN)_2^+$, ein Mittel zum Studium homogener Reaktionen des einwertigen Kupfers in wässriger Lösung. Experientia 19:488–489
23. Bissig K-D, Voegelin TC, Solioz M (2001) Tetrathiomolybdate inhibition of the *Enterococcus hirae* CopB copper ATPase. FEBS Lett 507:367–370
24. Brewer GJ, Askari F, Dick RB et al (2009) Treatment of Wilson's disease with tetrathiomolybdate: V. Control of free copper by tetrathiomolybdate and a comparison with trientine. Transl Res 154:70–77

25. Bell PF, Chen Y, Potts WE et al (1991) A reevaluation of the Fe(III), Ca(II), Zn(II), and proton formation constants of 4,7-diphenyl-1,10-phenanthrolinedisulfonate. Biol Trace Elem Res 30:125–144

26. McPhail DB, Goodman BA (1984) Tris buffer—a case for caution in its use in copper-containing systems. Biochem J 221:559–560

Chapter 2
Copper Toxicity

Abstract Copper is essential for life, yet highly reactive and a potential source of cell damage. Therefore, all cells possess copper homeostatic mechanisms to keep intracellular copper at safe levels. However, under conditions of excess environmental copper, homeostatic system become overloaded and intracellular copper rises to toxic levels. Possible toxic effects of copper span a range of mechanisms and it cannot be known with certainty which mechanism is active to what extent in a particular bacterium of vast and varied bacterial world. For common laboratory species like *Escherichia* or *Bacillus*, the concept has emerged that the main toxic action of copper is the replacement of iron in iron-sulfur cluster proteins, thereby inactivating essential enzyme functions.

Keywords Hydroxyl radical · Hydrogen peroxide · Fenton · Glutathione
Iron-sulfur cluster · Thiol depletion

The mechanism whereby copper is toxic to cells has generally been ascribed to the redox properties of copper, resulting in lethal oxidative damage to cells. However, recent work has put this concept into question and it is currently believed that the main toxic action of copper is the replacement of the iron cofactor in iron-sulfur cluster proteins. Other toxicity mechanism may still be at work to various extents, so to say in the background, depending on environmental and growth conditions. All major possible toxicity mechanisms will be described in this chapter and are summarized in Fig. 2.1. It must be stressed that the toxicity mechanisms discussed in this chapter do not apply to the antimicrobial action of metallic copper surfaces (see Sect. 2.4).

© The Author(s) 2018
M. Solioz, *Copper and Bacteria*, SpringerBriefs in Biometals,
https://doi.org/10.1007/978-3-319-94439-5_2

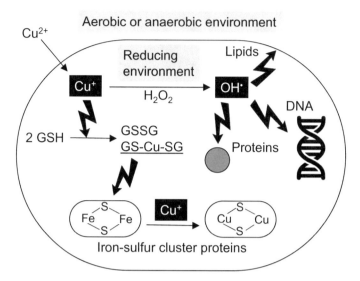

Fig. 2.1 Mechanisms of copper toxicity. Copper enters the bacterial cell by unknown pathways. The reducing condition in the cytoplasm reduces the copper to Cu^+, which can then participate in Fenton-type reactions to produce highly reactive hydroxyl radicals. These can in turn react nonspecifically with lipids, proteins and nucleic acid. Cu^+ can also lead to thiol depletion in the GSH pool, but also in proteins and free amino acids. Under anaerobic conditions, glutathione-copper complexes (GS–Cu–SG) can act as copper-donors for metalloenzymes. The dominant toxicity mechanism most likely is the displacement of iron from iron-sulfur cluster proteins by Cu^+.

2.1 Copper Toxicity Through the Formation of Reactive Oxygen Species

Copper can participate in a number of chemical reactions which lead to the generation of reactive oxygen species (ROS). Reactive hydroxyl radicals, which are extremely reactive in the cellular context, can be generated by a Fenton-type reaction (1):

$$Cu^+ + H_2O_2 = Cu^{2+} + OH^- + OH^{\cdot} \tag{1}$$

ROS production can be amplified by a combination with the Haber–Weiss cycle [(2) and (3)].

$$H_2O_2 + OH^{\cdot} = H_2O + O_2^- + H^+ \tag{2}$$

$$H_2O_2 + O_2^- = O_2 + OH^- + OH^{\cdot} \tag{3}$$

This could provide a particularly rich source of ROS in lactic acid bacteria which can accumulate large amounts of hydrogen peroxide [1, 2]. The rate constant of reaction (3) by itself is negligible, but Cu^{2+} or Fe^{3+} complexes can act as catalysts. Irreversible cell damage by ROS, particularly by the extremely reactive hydroxy radicals, can come about by a variety of mechanisms, such as inhibition of respiration, lipid peroxidation, or oxidative damage of proteins [3, 4].

Copper can also lead to depletion of glutathione (GSH), which is a major protective substance against heavy metal toxicity (see Sect. 2.3). This could occur in a cycle between reactions (4) and (5):

$$2Cu^+ + 2H^+ + O_2 = 2Cu^{2+} + H_2O_2 \qquad (4)$$

$$2Cu^{2+} + 2GSH = 2Cu^+ + GSSG + 2H^+ \qquad (5)$$

These combined reactions catalyze redox cycling of copper at the expense of GSH and O_2 to produce GSSG, the oxidized, dimeric form of GSH. Other cellular thiols could be depleted by the a similar mechanism. Hydrogen peroxide generated by reaction (4) could in turn participate in reactions (1)–(3) and amplify toxic hydroxyl radical production. While reasonable on paper, it is not clear if these reactions really occur in the cytoplasm under copper stress. The free copper concentration is probably far too low to catalyze these reactions.

The concept of cellular damage by copper *via* the production of ROS, thiol depletion, and oxidative damage of proteins, lipids, and DNA appears logical and has for years been claimed to be the toxicity mechanism of copper, but has never thoroughly been proven. Clearly, detrimental ROS production and oxidative damage can lead to cell death under certain stress conditions. For example, Woodmansee et al. [5] showed that nitric oxide (NO) accelerated the rate at which H_2O_2 killed *Escherichia coli* cells, apparently by greatly enhancing DNA damage through Fenton chemistry. Since NO damages the iron-sulfur clusters of dehydratases, the released iron could catalyze the Fenton reaction. However, NO also blocks respiration, which makes cells more susceptible to oxidative damage, making the mechanism of cell death unclear.

In the analysis of copper toxicity in *E. coli*, it was observed that the majority of H_2O_2-oxidizable copper is located in the periplasm and copper-mediated hydroxyl radical formation mainly occurs in this compartment, away from the DNA [6]. Copper-loading of cells actually increased their resistance to killing by H_2O_2 by eliminating iron-mediated oxidative killing and reducing the rate of DNA damage. These observations do not explain how copper suppresses iron-mediated damage but it is clear that copper does not catalyze significant oxidative DNA damage *in vivo*; therefore, copper toxicity must occur by a different mechanism. This challenges the oxidative-damage copper toxicity concept and newer work strongly supports a mechanism whereby the main toxic action of copper is the displacement of iron by copper in essential iron-cluster enzymes, thereby rendering key enzymes inactive.

2.2 Copper Toxicity by Iron-Sulfur Cluster Damage

First, it was demonstrated that the intracellular free concentration of copper, the Fenton reagent in the oxidative damage concept, is in the zeptomolar (10^{-21} M) range [7]. Intracellular copper, which appears to always be in the Cu^+ form, binds to sulfhydryl-bearing proteins, amino acids, and small molecules like glutathione (GSH). This, in combination with the copper-homeostatic machinery which pumps copper out of the cell, keeps the intracellular concentration of water-coordinated and thus reactive copper vanishingly small. Secondly, it was convincingly shown for *Escherichia coli* that the primary cause of copper toxicity is not the generation of hydroxyl radicals, but the displacement of iron by copper from the iron-sulfur clusters ([4Fe–4S] clusters) of important enzymes [8]. This is a thermodynamically favorable reaction because Cu^+ is a softer Pearson acid than Fe^{2+} and can thus effectively compete with iron for the cysteine ligands of iron-sulfur clusters (see Chap. 1). Primary targets of copper-iron exchange were dihydroxy-acid dehydratase of branched-chain amino acid synthesis, isopropylmalate isomerase involved in leucine biosynthesis, and fumarase A. The toxicity mechanism did not appear to be influenced by aerobic or anaerobic growth conditions.

Similar damage to iron-sulfur clusters was also demonstrated in the Gram-positive bacterium *Bacillus subtilis*. It was found that Cu^+ damaged the iron-sulfur cluster of SufU, which serves as the major scaffold for iron-sulfur cluster assembly and transfer to target proteins [9]. No significant induction of the PerR regulon, involved in oxidative stress defense, was observed under these conditions, indicating that oxidative stress did not play a major role. In *E. coli* grown under anaerobic conditions and amino acid limitation, but without exogenously added copper, endogenously liberated copper ions were found to damage the iron-cluster enzyme fumarate reductase and iron-cluster biogenesis [10]. These growth conditions also led to the induction of the alternative Suf system for iron cluster biogenesis and the CusCFBA copper transporter which pumps copper out of the periplasmic space (see Chap. 4). In *Rubrivivax gelatinosus* defective in the cytoplasmic copper exporter, CopA, and grown under microaerobic or anaerobic conditions in the presence of copper, a substantial decrease of cytochrome *c* oxidase and the photosystem was observed [11]. This led to reduced cytochrome oxidase and photosystem biogenesis, but also to coproporphyrin III extrusion from cells.

All these studies are in support of the newly emerged concept that copper toxicity is primarily due the poisoning of iron-sulfur cluster proteins by copper rather than oxidative damage. This raises questions about the applicability of the well-established in vitro Fenton redox processes of aqueous copper to the physiological regime. The concept of iron cofactor-displacement is further supported by the toxicity mechanisms of other heavy metal ions. Cd^{2+}, Ag^+, Zn^+, and Hg^{2+} have in common with copper that they have a soft Pearson character (high thiophilicity [12]), and thus also efficiently compete iron out of iron-sulfur clusters [13]. Cd^{2+}, Ag^+, and Zn^+ are not redox active and can thus not catalyze Fenton-type reactions. In line with this copper toxicity concepts, Park et al. showed that intracellular hydroxyl radical levels are not significantly changed by the addition of Cu^{2+} to *E. coli* [14]. Rather, the biocidal action of Cu^{2+} is attributable to the cytotoxicity of cellularly generated Cu^+, which does not appear to be associated with oxidative damage by Cu(I)-driven ROS. Cu^+ is considerably more toxic to cells than Cu^{2+} due to its higher thiophilicity and thus higher avidity for sulfhydryl residues of proteins, but also the higher permeability of the cytoplasmic membrane for Cu^+ than Cu^{2+} [15].

The concept of metal toxicity by replacement of iron in iron-sulfur clusters is further supported by the finding that cobalt stress also affects the function of iron-sulfur cluster proteins. Exposure of *E. coli* to cobalt resulted in the inactivation of the three iron-sulfur cluster enzymes tRNA methylthiotransferase, aconitase, and ferrichrome reductase. However, cobalt did not directly displace the iron from these protein. Rather, cobalt affected the iron-sulfur cluster assembly machinery *via* the scaffold proteins SufA and IscU, in which the iron-sulfur clusters are more labile [16]. Co^{2+}, being an intermediately-soft Pearson acid like Fe^{2+}, can successfully compete with iron for thiolate binding sites, particularly if it is present in excess.

In a transcriptomics study on Cd^{2+} toxicity in *E. coli*, the following genes/functions were observed to be upregulated: disulfide bond repair, oxidative damage repair, cysteine and iron-sulfur cluster biosynthesis, proteins with iron-sulfur clusters, iron storage proteins, and cadmium resistance proteins; general energy conservation pathways and iron uptake were down-regulated [17]. These findings are in line with the concept that Cd^{2+}, like Cu^+, effectively competes with iron for sulfur ligands. Released iron would then result in down-regulation of iron uptake and upregulation of iron storage proteins.

The concept of iron displacement from iron-sulfur cluster proteins does not preclude intracellular ROS generation. Indeed, displacement of iron from iron–sulfur clusters leads to increased cytoplasmic Fe^{2+}, which can catalyze Fenton chemistry. Copper ions induce the *soxRS* regulatory system of *E. coli* under aerobic conditions, indicating the generation of ROS, and this SOS response system can apparently cope with the resulting oxidative stress; hypersensitivity to copper is only observed in mutants deficient in superoxide dismutases or repair enzymes for oxidative DNA damage [18].

Possible alternative routes of copper toxicity include the occupation of zinc or other metal sites in proteins, and unspecific binding to proteins, lipids, and nucleic acids. Given the diversity of the bacterial community, there will be many variations to the scheme of copper toxicity, but the frequently stated copper-induced oxidative damage concept currently falls short of explaining bacterial copper toxicity. For an exhaustive discussion of metal ion toxicity in general, see Ref. [19].

2.3 Glutathione and Copper Toxicity

The small tri-peptide γ-L-glutamyl-L-cysteinylglycine, or glutathione (GSH), is present in the cytoplasm of all eukaryotes. In the prokaryotic world, GSH is absent in most Actinomycetes, which contain mycothiol instead [20]. GSH is present in many other Gram-positive bacteria and in most Gram-negative ones [21]. While most bacteria synthesize GSH in the cytoplasm, some take it up from the environment [22]. Some bacteria devoid of GSH contain bacillithiol or γ-L-glutamyl-L-cysteine instead [23, 24]. GSH is in equilibrium with its oxidized, dimeric form, GSSG, but GSH is the predominant form inside the cell. In *E. coli*, GSH/GSSG is the main redox couple that helps to maintain the cytoplasmic redox potential in growing cells at about -240 mV [25]. It is reasonable to assume that all microorganism contain a small-molecular weight thiol that can function similar to GSH.

GSH strongly complexes a variety of metal ions and has been shown to have a protective role in heavy metal stress in a number of organisms (see Chap. 3). However, recent work shows that under anaerobic conditions, GSH enhances copper toxicity in *Lactococcus lactis*. To specifically address detoxification independent of ROS formation, Obeid et al. [26] used an *L. lactis* strain which could neither synthesize GSH nor import it from the culture medium. GSH synthesis could be activated in this strain by inducing GSH biosynthetic enzymes from a plasmid [27]. Under anaerobic, fermentative conditions, GSH rendered *L. lactis* more sensitive to copper, particularly during the first phase of exponential growth [26]. It was proposed that GSH binds copper and facilitates its delivery to metal binding sites of enzymes, such as to iron-sulfur clusters. Such a mechanism would further support the hypothesis that metal sites of enzymes are the primary target of copper toxicity, as described in Sect. 2.2. That GSH-metal ion complexes facilitate the metallation of enzymes has been well documented in vitro [28–30]. These findings differ from those obtained under aerobic conditions, where GSH has been shown to exert a protective effect against copper toxicity both, in Gram-positive and Gram-negative organisms (see Chap. 3).

2.4   Copper Toxicity in Contact Killing

Bacteria are killed on dry copper surfaces in minutes to hours, a process referred to as contact killing [31]. The process has received considerable attention in recent years because copper could be used for critical touch-surfaces to curb the spread of infections [32]. There have been several hospital studies and the results look promising, but further work is needed [33–36]: Other applications of copper that appears attractive include the coating of surfaces of implants to avoid periprosthetic infections [37] or the use of copper-impregnated fabrics in health care settings [38].

There is now substantial insight into the mechanism of contact killing and it has become clear that the contact-killing process follows principles different from those in killing or growth arrest of bacteria by copper ions in suspension or in culture. Contact-killing as it is understood today proceeds as follows [31, 39, 40]:

(1) Copper dissolves from the copper surface and mM concentrations accumulate in the limited aqueous space within minutes.
(2) Severe membrane damage occurs and the cytoplasm is flooded with copper ions.

(3) Cytoplasmic content is lost and copper inhibits most metabolic activities.
(4) Genomic and plasmid DNA gets degraded by unknown mechanisms.
(5) No structurally intact bacteria can usually be detected after exposure to metallic copper.

These dramatic changes are in stark contrast to inhibition experiments in culture, where the bacteria are under growth conditions and thus have an energy supply to combat copper entry and repair some cell damage. Also, cells exposed to copper in culture do not appear to structurally disintegrate.

Finally, the meaning of "copper concentration" must be considered. In contact killing, there are usually no copper-binding growth media components around, so the copper liberated from a copper surface will lead to very high 'free' copper concentrations. Copper avidly binds to growth media components, so the 'free' or active copper concentration in *culture* is not anywhere near to the 'added' copper concentration. Estimates from our lab suggest that if 5 mM CuSO$_4$ is added to standard LB growth media, the free copper concentration is <10 μM, and is further lowered with increasing cell density (M. Solioz, unpublished observation). There are two corollaries to this: first, it is virtually impossible to compare copper 'concentrations' between labs and, secondly, it is impossible to compare copper concentrations between contact killing experiments and inhibition experiments in culture.

References

1. Baureder M, Reimann R, Hederstedt L (2012) Contribution of catalase to hydrogen peroxide resistance in *Enterococcus faecalis*. FEMS Microbiol Lett 331:160–164
2. van de Guchte M, Serror P, Chervaux C et al (2002) Stress responses in lactic acid bacteria. Antonie Van Leeuwenhoek 82:187–216
3. Nandakumar R, Espirito Santo C, Madayiputhiya N et al (2011) Quantitative proteomic profiling of the *Escherichia coli* response to metallic copper surfaces. Biometals 24:429–444
4. Yoshida Y, Furuta S, Niki E (1993) Effects of metal chelating agents on the oxidation of lipids induced by copper and iron. Biochim Biophys Acta 1210:81–88
5. Woodmansee AN, Imlay JA (2003) A mechanism by which nitric oxide accelerates the rate of oxidative DNA damage in *Escherichia coli*. Mol Microbiol 49:11–22
6. Macomber L, Rensing C, Imlay JA (2007) Intracellular copper does not catalyze the formation of oxidative DNA damage in *Escherichia coli*. J Bacteriol 189:1616–1626
7. Changela A, Chen K, Xue Y et al (2003) Molecular basis of metal-ion selectivity and zeptomolar sensitivity by CueR. Science 301:1383–1387
8. Macomber L, Imlay JA (2009) The iron-sulfur clusters of dehydratases are primary intracellular targets of copper toxicity. Proc Natl Acad Sci USA 106:8344–8349
9. Chillappagari S, Seubert A, Trip H et al (2010) Copper stress affects iron homeostasis by destabilizing iron-sulfur cluster formation in *Bacillus subtilis*. J Bacteriol 192:2512–2524
10. Fung DK, Lau WY, Chan WT et al (2013) Copper efflux is induced during anaerobic amino acid limitation in *Escherichia coli* to protect iron-sulfur cluster enzymes and its biogenesis. J Bacteriol 195:4556–4568
11. Azzouzi A, Steunou AS, Durand A et al (2013) Coproporphyrin III excretion identifies the anaerobic coproporphyrinogen III oxidase HemN as a copper target in the Cu-ATPase mutant copA of *Rubrivivax gelatinosus*. Mol Microbiol 88:339–351

12. Pearson RG (1968) Hard and soft acid and bases, HSAB, part I. J Chem Educ 45:581–587
13. Xu FF, Imlay JA (2012) Silver(I), mercury(II), cadmium(II), and zinc(II) target exposed enzymic iron-sulfur clusters when they toxify *Escherichia coli*. Appl Environ Microbiol 78:3614–3621
14. Park HJ, Nguyen TT, Yoon J et al (2012) Role of reactive oxygen species in *Escherichia coli* inactivation by cupric ion. Environ Sci Technol 46:11299–11304
15. Abicht HK, Gonskikh Y, Gerber SD et al (2013) Non-enzymatic copper reduction by menaquinone enhances copper toxicity in *Lactococcus lactis* IL1403. Microbiol 159:1190–1197
16. Ranquet C, Ollagnier-de-Choudens S, Loiseau L et al (2007) Cobalt stress in *Escherichia coli*. The effect on the iron-sulfur proteins. J Biol Chem 282:30442–30451
17. Helbig K, Grosse C, Nies DH (2008) Cadmium toxicity in glutathione mutants of *Escherichia coli*. J Bacteriol 190:5439–5454
18. Kimura T, Nishioka H (1997) Intracellular generation of superoxide by copper sulphate in *Escherichia coli*. Mutat Res 389:237–242
19. Lemire JA, Harrison JJ, Turner RJ (2013) Antimicrobial activity of metals: mechanisms, molecular targets and applications. Nat Rev Microbiol 11:371–384
20. Newton GL, Arnold K, Price MS et al (1996) Distribution of thiols in microorganisms: mycothiol is a major thiol in most actinomycetes. J Bacteriol 178:1990–1995
21. Fahey RC, Brown WC, Adams WB et al (1978) Occurrence of glutathione in bacteria. J Bacteriol 133:1126–1129
22. Li Y, Hugenholtz J, Abee T et al (2003) Glutathione protects *Lactococcus lactis* against oxidative stress. Appl Environ Microbiol 69:5739–5745
23. Gaballa A, Newton GL, Antelmann H et al (2010) Biosynthesis and functions of bacillithiol, a major low-molecular-weight thiol in Bacilli. Proc Natl Acad Sci USA 107:6482–6486
24. Kim EK, Cha CJ, Cho YJ et al (2008) Synthesis of γ-glutamylcysteine as a major low-molecular-weight thiol in lactic acid bacteria *Leuconostoc* spp. Biochem Biophys Res Commun 369:1047–1051
25. Schafer FQ, Buettner GR (2001) Redox environment of the cell as viewed through the redox state of the glutathione disulfide/glutathione couple. Free Radic Biol Med 30:1191–1212
26. Obeid MH, Oertel J, Solioz M et al (2016) Mechanism of attenuation of uranyl toxicity by glutathione in *Lactococcus lactis*. Appl Environ Microbiol 82:3563–3571
27. Fu RY, Bongers RS, van Swam II et al (2006) Introducing glutathione biosynthetic capability into *Lactococcus lactis* subsp. *cremoris* NZ9000 improves the oxidative-stress resistance of the host. Metab Eng 8:662–671
28. Musci G, Di Marco S, Bellenchi GC et al (1996) Reconstitution of ceruloplasmin by the Cu(I)-glutathione complex. Evidence for a role of Mg^{2+} and ATP. J Biol Chem 271:1972–1978
29. Ferreira AM, Ciriolo MR, Marcocci L et al (1993) Copper(I) transfer into metallothionein mediated by glutathione. Biochem J 292:673–676
30. Ciriolo MR, Desideri A, Paci M et al (1990) Reconstitution of Cu, Zn-superoxide dismutase by the Cu(I).glutathione complex. J Biol Chem 265:11030–11034
31. Grass G, Rensing C, Solioz M (2011) Metallic copper as an antimicrobial surface. Appl Environ Microbiol 77:1541–1547
32. Grass G, Hans M, Mücklich F et al (2015) Massive Kupferwerkstoffe in der Hygiene und Infektionsprävention - Zu gut um wahr zu sein? Hyg Med 40:458–463
33. Sifri CD, Burke GH, Enfield KB (2016) Reduced health care-associated infections in an acute care community hospital using a combination of self-disinfecting copper-impregnated composite hard surfaces and linens. Am J Infect Control 44:1565–1571
34. Schmidt MG, von Dessauer B, Benavente C et al (2016) Copper surfaces are associated with significantly lower concentrations of bacteria on selected surfaces within a pediatric intensive care unit. Am J Infect Control 44:203–209
35. Michels HT, Keevil CW, Salgado CD et al (2015) From laboratory research to a clinical trial: Copper alloy surfaces kill bacteria and reduce hospital-acquired infections. Health Environ Res Des J 9:64–79

36. Salgado CD, Sepkowitz KA, John JF et al (2013) Copper surfaces reduce the rate of healthcare-acquired infections in the intensive care unit. Infect Control Hosp Epidemiol 34:479–486
37. Norambuena GA, Patel R, Karau M et al (2017) Antibacterial and biocompatible titanium-copper oxide coating may be a potential strategy to reduce periprosthetic infection: an *in vitro* study. Clin Orthop Relat Res 475:722–732
38. Humphreys H (2014) Self-disinfecting and microbiocide-impregnated surfaces and fabrics. What potential in interrupting the spread of healthcare-associated infection? Clin Infect Dis 58:848–853
39. Hans M, Mathews S, Mücklich F et al (2016) Physicochemical properties of copper important for its antibacterial activity and development of a unified model. Biointerphases 11:018902-1–018902-8
40. Luo J, Hein C, Mücklich F et al (2017) Killing of bacteria by copper, cadmium, and silver surfaces delineates the relevant physico-chemical parameters. Biointerphases 12:020301-1–020301-6

Chapter 3
Copper Homeostasis in Gram-Positive Bacteria

Abstract Copper homeostasis in Gram-positive bacteria essentially requires three components: a copper exporting ATPase (founding member CopA), a copper chaperone (founding member CopZ), and a copper-responsive regulator (founding member CopY) which regulates the expression of these functions. GSH works as a back-up protective system. These four components are also part of the more complex copper homeostatic mechanism of Gram-negative organisms, discussed in Chap. 4, but will only be discussed in detail in this Chapter. Copper reduction at the plasma membrane, which is specific to Gram-positive bacteria will also be discussed here. Finally, copper regulons, which are apparently specific to Firmicutes, are presented. Most of the findings described here derive from *Enterococcus hirae* and *L. lactis*, but other organisms are considered as necessary.

Keywords ATPase · Chaperone · Repressor · CopA · CopZ · CopY Gram-positive · *Lactococcus* · *Enterococcus* · Regulon

The first copper ATPases to be cloned were CopA and CopB of *Enterococcus hirae* [1, 2] and this organism has stood at the forefront of bacterial copper homeostasis research for many years. Thus, bacterial copper homeostasis is probably best understood in this organism and it serves as a paradigm for copper homeostasis in Gram-positive bacteria. There are three core element of copper homeostasis that are present in all bacteria, Gram-negative or Gram-positive: a copper exporting ATPase (CopA), a copper chaperone (CopZ), and a copper-responsive transcriptional regulator which (CopY). CopY typically regulates its own expression and that of CopA and CopZ and these three genes are arranged in an operon. There are many different possible arrangements of the core copper homeostasis genes, in line with the observation that operon structures are less conserved than gene functions [3]. Figure 3.1 shows a few operon structures encountered in Gram-positive bacteria to illustrate this point. There exist many more divergent operon arrangements.

These and other elements are summarized and numbered in Fig. 3.2 and will be discussed chronologically in detail below, with Chapter subheadings correspond to the numbers in Fig. 3.2. The copper regulons, which are discussed at the end of this chapter, do not appear in Fig. 3.2.

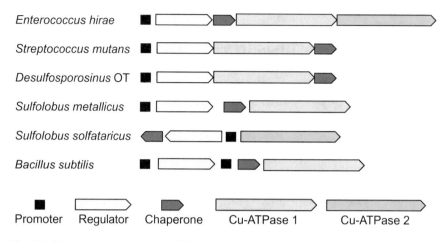

Fig. 3.1 Selected operon structures of Gram-positive bacteria

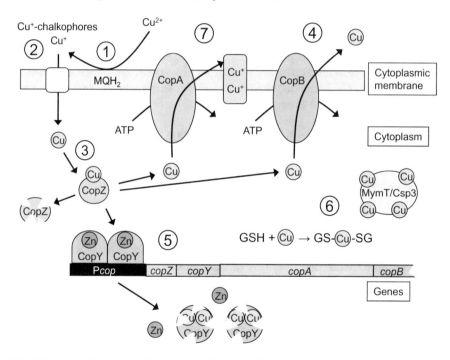

Fig. 3.2 Schematic representation of copper homeostasis in Gram-positive bacteria. The circled numbers correspond to the subheadings of this chapter: 1, extracellular copper reduction; 2, copper entry into the cytoplasm; 3, copper sequestration by metallochaperones; 4, copper secretion by ATPases; 5, copper-activation of gene expression; 6, copper binding by glutathione and proteins. See text for details

3.1 Extracellular Copper Reduction

Microorganisms can exchange electrons with inorganic ions in solution or in minerals in a variety of way and use them as electron donors as well as electron acceptors (see Ref. [4] for review). *Lactococcus lactis* IL1403, a member of the Gram-positive Firmicutes like *E. hirae* or Bacilli, was found to possess a pronounced extracellular Cu^{2+}-reductase activity. This results in the accumulation of Cu^+ outside the cell. Since Cu^+ is more toxic than Cu^{2+}, copper reduction works as a suicide mechanism and inhibits cell growth. Copper reduction was shown to be directly catalyzed by menaquinones in the membrane without the involvement of an enzyme [5]. The electrons are transferred from NADH via membrane-bound quinones to Cu^{2+} (Fig. 3.3a). Driving force is the large potential difference between the NADH/NAD$^+$ and the Cu^{2+}/Cu$^+$ redox couples of +465 mV. Facultative anaerobes like *L. lactis* generally produce an excess of reducing equivalents (or electrons) when growing fermentatively and extracellular copper reduction provides an opportunity to dispose of surplus reducing equivalents. When grown in milk, *L. lactis* similarly disposes of reducing equivalents by reducing the redox potential of milk from +300 to −220 mV [6].

Some facultative anaerobes like *L. lactis* IL1403 can express a terminal cytochrome oxidase. They harbor the genes for the two subunits of a *bd*-type terminal cytochrome oxidase, but are unable to synthesized heme. So when provided with heme, the organisms synthesize a functional cytochrome *bd*-oxidase and turn into respiring aerobes, accompanied by all the features of aerobic bacteria, like a dramatic increase in growth yield, insensitivity to oxygen, etc. [7]. However, when expressing *bd*-cytochrome oxidase, extracellular copper reduction is abolished. It was shown by Abicht et al. [5] that this due to competition for electrons between oxygen and Cu^{2+} (Fig. 3.3b). The potential difference between the NADH/NAD$^+$ and the Cu^{2+}/Cu$^+$ couples is +465 mV, while that between the NADH/NAD$^+$ and the O_2/H$_2$O couple is +1.5 V.

Most likely, many other Firmicutes are able to catalyze extracellular copper reduction and some of these organisms will also be able to express a functional terminal oxidase under the right conditions. These aspects have, to the knowledge of the author, never been systematically addressed.

3.2 Copper Entry into the Cytoplasm

How copper enters bacteria has remained unknown for many years, but new concepts are starting to emerge. A widely held opinion was, that there is no need for cytoplasmic copper in bacteria because all known bacterial cuproenzymes are localized either in the cytoplasmic membrane or the periplasm. The assembly of cupro-enzymes can thus take place on the outer face of the cytoplasmic membrane in Gram-positive bacteria or in the periplasmic space in Gram-negative bacteria. However, it is now becoming clear that for cuproenzyme assembly, copper usually enters the cytoplasm

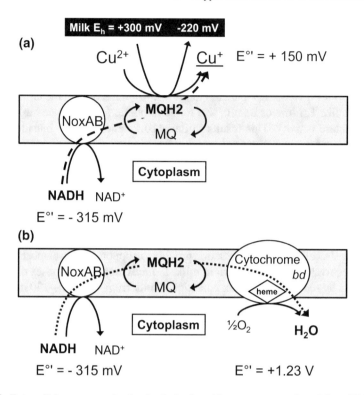

Fig. 3.3 Extracellular copper reduction by *L. lactis*. **a** Electrons are transferred from NADH via NoxAB NADH oxidase to menaquinones in the membrane and, in a non-enzymatic reaction, to extracellular Cu^{2+}. The overall flow of electrons is indicated by the dashed arrow. **b** In *L. lactis* grown in the presence of heme and thus expressing cytochrome *bd* oxidase, electrons flow from NADH via NoxAB NADH oxidase to menaquinones in the membrane, and then via cytochrome *bd* oxidase to oxygen. The overall flow of electrons is indicated by the dotted arrow

first and is then re-exported by specific 'assembly' exporters for enzyme metallation (discussed in detail in Chap. 4). This insures high cofactor–specificity in the metallation of process. Copper from the environment can also access the outer face of the cytoplasmic membrane, such as through porins and without the need for specific transport systems. In some rare cases (e.g. CueO), this copper can participates in the metallation of cuproenzymes. But it begins to emerge that specific copper uptake systems for cytoplasmic copper exist (see also Chap. 4). It also appears that some copper enters the bacterial cytoplasm inadvertently, either as a 'blind passenger' bound to organic molecules or via transporters for inorganic ions, such as phosphate transporters. Copper ions can probably also penetrate the lipid bilayer: it was found that the more membrane-permeable Cu^+ ions are more toxic than the less membrane-permeable Cu^{2+} ions [5]. These non-specific copper supply routes maybe sufficient under some, but probably not all, environmental conditions.

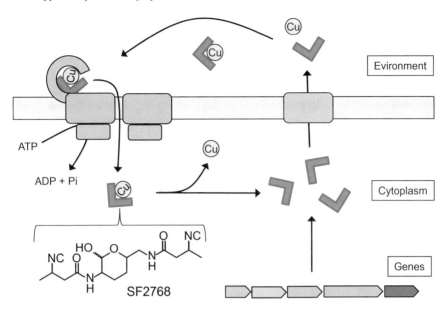

Fig. 3.4 Ligand-guided copper uptake by Gram-positive organisms. The chalkophore (blue angle), synthesized either ribosomally or non-ribosomally, is secreted into the extracellular space by an export protein (green). Cu^+ or Cu^{2+} is scavenged from the environment by the chalkophore and the Cu-chalkophore complex is actively taken up by the cell, such as by a chelate ABC-transporter or a similar system (brown) in an ATP–dependent manner. In the cytoplasm, Cu^+ is released from the chalkophore, and the *apo*-chalkophore is recycled. The structure of the chalkophore SF2668 is also shown

There was also misleading information about copper uptake by *E. hirae* in the literature: it had been claimed that the CopA copper ATPases is for copper import, while the CopB copper ATPase is for copper secretion [8]. This interpretation was later revised—*E. hirae* has no known copper uptake system, but this has not always found its way into the more recent literature [9]. Accordingly, these copper defense systems only serve in the removal of cytoplasmic copper that has entered the cell inadvertently. There have also been of some reports of copper acquisition by P1B-type copper ATPases in Gram-negative bacteria (e.g. [10–12]). These and other reports on ATPase-driven copper uptake were based on indirect observations, never direct copper transport, and were single reports without any follow-up work. Current evidence supports the view that all P1B-type copper ATPases catalyze copper export.

Other systems for copper entry into Gram-positive bacteria have been proposed. In *Bacillus subtilis*, YcnJ has been claimed to be a copper importer [13], but a role in cytochrome oxidase assembly appears also possible. Also, ZosA of *B. subtilis* has been suggested to be a copper importer, but no direct transport evidence yet exists [14]. In the same organism, it was also found that the *ycnKJI* operon is induced by copper limitation and the YcnJ membrane protein was proposed to catalyze copper uptake by the cells. But here too, further proof is required [15].

It was found that Gram-negative methanotrophs synthesize methanobactins, which are chalkophores, or a 'siderophore' for copper (Greek 'chalko' = copper), secreted in response to copper limitation (see Chap. 4). A similar such ligand was identified in the Gram-positive Actinomycetes *Streptomyces thioluteus* [16]. In this organism, a non-ribosomal peptide synthetase gene cluster directs the biosynthesis of the small-molecular weight natural product SF2768. An ATP-binding cassette (ABC) transporter related to iron import within the biosynthesis gene cluster for SF2768 suggested that the product might be a siderophore (Fig. 3.4). However, characterization of the metal-binding properties of SF2768 revealed that it specifically binds copper, not iron, and is thus a chalkophore. Indeed, the intracellular copper content of *S. thioluteus* increased upon incubation with copper-SF2768, showing copper acquisition via SF2768 as a mechanism for copper uptake by *S. thioluteus*. Chalkophore-assisted copper uptake clearly is a widespread mechanism that yet awaits discovery in other organisms.

3.3 Copper Chaperones

Copper chaperones or, more generally, metallochaperones, are terms coined and officially registered in the late nineteen-nineties to describe small proteins which are involved in the handling of metal ions. The term 'chaperone' is unfortunately ambiguous because the word 'chaperone' is also used for proteins which aid in proper folding of nascent polypeptide chains, so 'metallochaperone' is preferable. There is a growing number of metallochaperones, not only for copper, but also for mercury, silver, nickel, etc. but the focus here rests on copper chaperones. One type of cytoplasmic copper chaperones, the CopZ-like chaperones, named after its founding member CopZ of *E. hirae*, is conserved across all phyla. They are small, 7–8 kDa, proteins which bind cytoplasmic copper via a conserved CXXC motif and deliver it to points of utilization or transport [17]. They also minimize copper toxicity by scavenging cytoplasmic copper and maintaining it in a tightly bound form. Other metallochaperones bind zinc, mercury, cobalt, iron, or nickel, and in Gram-negative bacteria also can have a periplasmic location (see Chap. 4).

The first copper chaperones were identified in yeast and bacteria in 1995. The yeast chaperone, ATX1, was identified as a protein that protects cells against the toxicity of superoxide anions and hydrogen peroxide [18] while CopZ of *E. hirae* was identified as a protein required for induction of the copper resistance operon [19]. Bioinformatics analysis showed that ATX1 and CopZ both resemble MerP, which functions in mercury resistance. Bacteria deal with toxic mercury in an unexpected way: Hg^{2+} is first transported into the cytoplasm by MerT, where it is reduced to Hg^0 by the MerA mercuric reductase. Volatile Hg^0 then leaves the cell by diffusion [20]. MerP is a periplasmic mercury chaperone required for Hg^{2+} transport into the cell by MerT [21]. This peculiar mercury resistance mechanism is mentioned here because it resembles copper-loading of enzymes, where copper also has to pass through the cytoplasm, as discussed in Chap. 4.

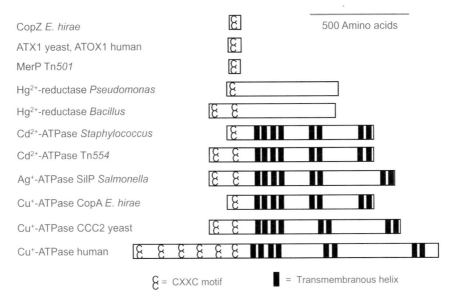

Fig. 3.5 Cartoon of the occurrence of CopZ-like metal binding domains in a range of proteins involved in heavy metal detoxification. The proteins are drawn to scale

Chaperone-like sequence elements are also present in variable numbers in the N-termini of heavy metal ATPases as well as in soluble mercuric reductases (Fig. 3.5). It is still a matter of debate what the function of metallochaperone-like domains fulfil in the N-termini of larger protein, particularly since these domains are often dispensable and also do not affect the ion specificity of the enzymes (see Sect. 3.4.) [22–24]. In contrast, the function of the small metallochaperones in metal ion-routing is now well understood.

CopZ of *E. hirae* was the first copper chaperone for which a structure was available [25]. It consists of a ferredoxin-like $\beta\alpha\beta\beta\alpha\beta$-fold; the CXXC metal-binding motif is located in a flexible loop connecting the first β-strand with the first α-helix (Fig. 3.6). Today, dozens of structures of metallochaperones and the related heavy metal binding domains of larger proteins are available and they all exhibit the same $\beta\alpha\beta\beta\alpha\beta$-fold [26–36]. The universality of this building block raises interesting questions about its evolution. An analysis in this regard supports a concept whereby copper chaperones (and probably chaperones for other heavy metals) and MBDs of larger enzymes represent ancient and distinct lineages that have evolved largely independently [37]. However, the problem probably requires re-appraisal with today's larger database and the better knowledge of chaperone function.

There is still debate of how copper is complexed by chaperones in the cytoplasm. The human CopZ-like copper chaperone (ATOX1, earlier called HAH1) was shown by X-ray crystallography to from a dimeric structure in which a single Hg^{2+} or Cu^+ atom is bonded by four cysteine residues [38]. Cu^+-complexation by CopZ

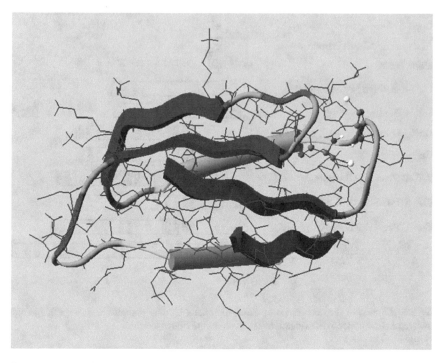

Fig. 3.6 Structure of the CopZ copper chaperone of *E. hirae*. The structure of the apo form of CopZ was derived by NMR spectroscopy and shows an βαββαβ thioredoxin fold. The two copper-binding cysteines, C12 and C15 (yellow), are situated in a mobile loop between the first β-strand and the first α-helix at the right end of the molecule (from structure 1CPZ of the PDB Protein Data Bank)

of *E. hirae* analyzed in solution by NMR also revealed a dimeric structure, but with trigonally bound copper to be the most likely structure [25]. Glutathione was shown to inhibit dimer formation in vitro and could, in principle, be a ligand to monomeric Cu^+-CopZ inside the cell, where glutathione concentrations are high. There are many other studies on chaperone-metal ion interaction with chaperones from various organisms, but the uncertainty of how Cu^+ is complexed by chaperones in vivo remains [39]. It was shown that CopZ can pick up Cu^+ from solution or from Cu^+-complexes like Cu^+-acetonitrile in vitro [25]. Copper which enters the cell will not be present as free Cu^+ ion, but will immediately be bound by proteins, glutathione, or other low–molecular weight sulfhydryls. Copper chaperones must be able to scavenge copper from these sites to channel it into the copper homeostatic system. What happens to chaperone-bound copper will be discussed in the context of the cognate target proteins in the following Sections.

Docking of CopZ-type chaperones to the N-terminal metal binding domains of Cu-ATPases has been studied in a variety of systems, eukaryotic as well as prokaryotic. It is clear that the two protein entities interact and that copper can be transferred from the chaperone to the ATPase [26, 28–31, 40–43]. However, the function of this copper transfer remains elusive; most likely, it has a regulatory function [44]. According to current evidence, MBDs are not the entry point of copper for transport by the ATPase (see Sect. 3.4).

The identification of copper chaperones marked the emergence of a new concept in the handling of metal ions by cells, namely, the escorting of the metal ion by a protein to prevent nonspecific, damaging interactions with other cell components. In addition to copper chaperones, we now know chaperones which function in the handling of nickel, mercury, cobalt, iron, and zinc, and we are still far from understanding all the intricacies of metal homeostasis and metalloprotein assembly. Eukaryotic organisms also possess a wide, but largely different set of specialized chaperones for the metallation of enzymes and to deliver copper to subcellular compartments (see Refs. [17, 45, 46] for further reading).

The extensive knowledge derived from in vitro experiments of how CopZ-type copper chaperones bind copper, interact with partner proteins, and transfer copper to other proteins contrast with the understanding of the in vivo function. Bacterial strains deficient of CopZ generally display no phenotype, or only display a phenotype under 'optimized' conditions. So CopZ appears dispensable, at least under laboratory conditions. There are indications all across bacterial copper research that laboratory conditions fall short of reflecting the 'real-life' situation. It will be important to direct more research at the investigation of bacteria in natural habitats and in bacterial communities. Modern tools and methodologies for this are now available.

3.4 Copper ATPases

3.4.1 Copper ATPase Structure

Copper ATPases belong to the superfamily of P-type ATPases. These enzymes are ATP-driven ion pumps embedded in the cytoplasmic membrane. Forty years ago, P-type ATPases were believed to be a typical eukaryotic feature, chiefly represented the Ca- and Na,K-ATPases, but it is now clear that they are common to all forms of life. Structurally, they consist of a major catalytic protein subunit of 70–200 kDa and transport a wide range of cations, ranging from H^+ to Hg^{2+}. Transport is driven by cytoplasmic ATP in a mechanism that is unique to these enzymes: the γ-phosphate of ATP is first transferred to the enzyme where it forms a covalent acyl-phosphate intermediate with the carboxyl group of an aspartic acid residue [47, 48]. Thereby, the energy of hydrolysis of ATP is not yet released directly, but transferred to the enzyme; it is this phosphorylated intermediate which has led to the name 'P-type' ATPases. In a later step, the phosphoester bond on the enzyme is broken to harness the energy for transport (cf. Fig. 3.10c). The phosphorylated aspartic acid is conserved across all P-type ATPases in the motif DKTGT (in the one-letter amino acid code, used throughout this book).

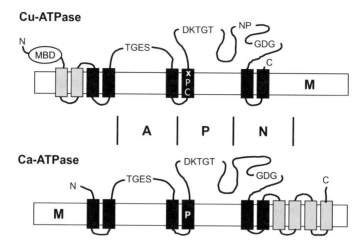

Fig. 3.7 Cartoon of the domain architecture of a bacterial Cu-ATPase and a eukaryotic Ca-ATPase. MBD, metal binding domain; M, membrane domain; A, actuator domain; P, phosphorylation domain; N, nucleotide binding domain; CPx, ion transduction site. Helices which differ between heavy metal and non-heavy metal ATPases are in grey. See text for conserved amino acid sequences indicated in the figure.

Initial structural work on the Ca-ATPase of the sarcoplasmic reticulum has revealed three distinct domains [49]. The transmembrane (M) domain encompasses 10 transmembrane helices (Fig. 3.7). A flexible actuator (A) domain characterized by the conserved motive TGES lies between membrane helices 2 and 3. The phosphorylation (P) domain harboring the DKTGT-phosphorylation site and the nucleotide binding (N) domain with a GDG signature connect membrane helices 4 and 5. Cu and most other heavy metal ATPases maintain the three cytoplasmic domains and the respective conserved sequences, but exhibit a different membrane helix architecture: they possess two additional N-terminal helices, but lack the four C-terminal helices of the Ca-ATPase and related enzymes (non-heavy metal ATPases). Several additional features are unique to Cu-ATPases and allows their identification on the basis of the amino acid sequence: (i) an intramembranous CPx-motif on helix 6 which serves as the ion gate and has led to the terminology 'CPx-type' ATPases, and (ii) a conserved NP-motif of unknown function in the N-domain, and (iii) one or several N-terminal metal binding domains (MBD, also called heavy metal-associated (HMA) domains).

The MBDs of Cu-ATPases can be of two types:

(1)   A domain of about 70 amino acids containing a conserved CxxC heavy metal binding motif [8, 50]. This is the most frequent type of MBD and forms a modular building block that can be repeated various times in the N-terminus of a Cu-ATPases, e.g. once in most bacterial Cu-ATPases, including CopA of *E. hirae*, twice in *E. coli* CopA and yeast CCC2, four times in fruit fly Cu-ATPase and six times in the human Cu-ATPases [28]. The reason for the variation in the number of MBDs and their precise function remain unclear.

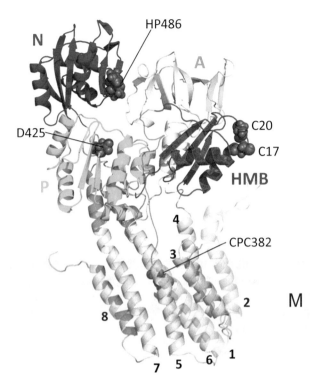

Fig. 3.8 Three-dimensional structure of CopB of *E. hirae* based on chemical crosslinking and modelling [52]. Indicated are the A, P, and N-domains, the intramembranous Cu transduction site, CPX382, the phosphorylation site, D425, the conserved HP465 motif, and the HMB domain with the conserved C17 and C20 residues. M indicates the membrane. See text for further explanations

(2) A very histidine-rich stretch of amino acids at the N-terminus. This type of N-terminal MBD is much rarer and has so far only been observed in bacteria. It is for example found in the N-terminus of CopB of *E. hirae*. The corresponding sequence is:

47-HHTHGHMERHQQMDHGHMSGMDHSHMDHE

DMSGMNHSHMGHENMSGMDHSMHM-100

This type of motif (in CopB of *E. hirae*, 15 histidines in 54 amino acids) has so far only been observed in single copies in Cu-ATPases. No functional differences are known between Cu-ATPases with CXXC type MBDs or histidine-rich MBDs.

Currently, only two structures of bacterial copper ATPases are available. An X-ray crystallographic structure of CopA from *Legionella pneumophila* shows that the overall structure resembles that of eukaryotic Ca-ATPase [51]. The core structures of the cytoplasmic domain are maintained, but are smaller than in the Ca-ATPase. The N-terminal HMB domain is not resolved in this structure due to its flexibility. It is however apparent in a structure derived by intramolecular crosslinking, followed by digestion with trypsin and identification of the crosslinked fragments by mass spectrometry (Fig. 3.8) [52]. Functional aspects of Cu-ATPases will be discussed in Sect. 3.4.3.

3.4.2 Evolution of P-Type ATPases

Over the years, the superfamily of P-type ATPases has grown substantially and now includes P-type ATPases for H^+, Na^+, K^+, Na^+K^+, K^+H^+, Mg^{2+}, Mn^{2+}, Fe^{2+}, Ni^{2+}, Ca^{2+}, Ag^+, Cu^+, Cd^{2+}, Zn^{2+}, Pb^{2+}, Co^{2+}, Au^+, and phospholipids (flippases) [53, 54]. Cu-ATPases belong to the family of P1B-type ATPases, which includes ATPases also transporting Ag^+, Cd^{2+}, Zn^{2+}, Pb^{2+}, and Co^{2+} (Fig. 3.9). The division of the P-type ATPase superfamily into families corresponds to the transport specificities rather than to the phylogeny of species, e.g. the Cu-ATPases of Gram-positive bacteria are closely related to those of mammalian cells, while the species could not be any more different. There are two corollaries to this: first, the specificities of P-type ATPases do not rest in a few key residues, but are an intrinsic property of the entire structure and, secondly, P-type ATPases most likely arose early in evolution, before the division of species into prokaryotes and eukaryotes.

In support of this, it has been found that *Methanococcus jannaschii* harbors a gene that encodes a predicted protein, MJ0968, of 273 amino acids with sequence similarity to the P-N domain of P-type ATPases, and thus covering all the amino acids involved in ATP binding and hydrolysis. When expressed in *E. coli*, MJ0968 can hydrolyze ATP, autophosphorylates, and is inhibited by orthovanadate, which is a typical feature of all P-type ATPases [56]. MJ0968 shares the highest sequence identity (30%) to the P1B-type Cu-ATPase of *Helicobacter pylori*. The soluble MJ0968 protein may thus be an ancestral P-type ATPase which functions in concert with other subunits; these subunits may later have become fused to a single-subunit P-type ATPase. There are many instances in nature where a multi-subunit protein also occurs in a 'fused' version. Finally, the defense against toxic heavy metal ions may have been a priority early in evolution, long before cells started to use copper as a bioelement. For such an early 'bilge pump', function, there was no need for gene regulation. Only once copper became a bioelement, regulatory systems had to evolve and diversify. This could explains the range of structurally unrelated copper-responsive regulators of gene expression we see today. So while the evolution of P-type ATPases started early and was divergent, that of copper-responsive transcriptional regulators started late and was convergent, by adapting existing regulators to copper (see Sect. 3.5).

The two *E. hirae* copper ATPases, CopA and CopB, marked the discovery of ATP-driven copper transport across cell membranes in 1992 (cf. Fig. 3.2) (Odermatt et al. [1]). It may be hard to comprehend this today but at that time, there was no concept of how copper crosses cell membranes. A favored concept invoked nonspecific penetration of the lipid bilayer by copper ions. The surprising discovery that *E. hirae* possesses not one, but two copper ATPases had led to the concept that one serves in copper import and the other in copper export [57]. Later work by others however showed

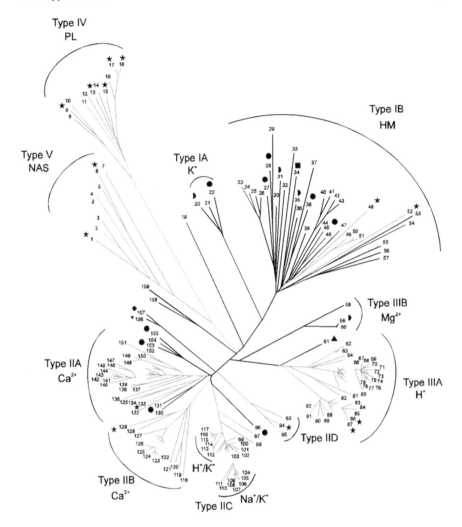

Fig. 3.9 Unrooted phylogenetic tree based on the core sequences of 159 P-type ATPases. Some areas which are not connected to the rest of the tree have been enlarged by 40% to clarify the distribution of species. When the substrate specificity of the ATPases present in each family is known, it corresponds in all cases to the name of the family. The *numbers* of the sequences correspond to the numbers defined in Table 1 of Ref. [55]. *Black branches,* ATPases of Bacteria and Archaea; *grey branches,* ATPases of Eukarya. The P-type ATPases from fully sequenced organisms are marked with the following symbols: ◗ *Escherichia coli*; ■ *Haemophilus influenzae*; ▲ *Methanococcus jannaschii*; ◆ *Mycoplasma genitalium*; ▼ *Mycoplasma pneumoniae*; ★ *Saccharomyces cerevisiae*; ● *Synechocystis* PCC6803. The transport specificities of the ATPases are indicated as follows: *HM,* heavy metals; *NAS,* no assigned specificity; *PL,* phospholipids. Reproduced with permission from Ref. [55]

that both enzymes catalyze copper export from the cytoplasm, albeit at different rates [58], and that one enzyme serves in the extrusion of excess cytoplasmic copper, while the other is required for the metallation of enzymes in the periplasmic space.

3.4.3   Copper ATPase Function

The reason for an organism to have two copper ATPase became clear from the study of *Pseudomonas aeruginosa*. It could be shown that CopA1 is required to maintain low cytoplasmic copper (for copper tolerance), while CopA2 is required to provide copper to the periplasm for cytochrome *c* oxidase metallation [59]. *E. hirae* or the related *Lactococcus lactis* IL1403 both feature two Cu-ATPases, of which only one could be shown to provide copper tolerance (CopB in *E. hirae*, CopA in *L. lactis*) [9]. The other Cu-ATPases do not convey copper tolerance and presumably serve in the metallation of cuproenzymes in the extracytoplasmic space (see Sect. 4.8.). No Cu-containing enzyme is known in either of these organisms, but it has to be kept in mind that (i) only about 70% of the genes of an organism have an assigned function, (ii) 70% of the assigned functions are wrong and (iii), these organisms are grown under unnatural conditions in the laboratory, limiting any biochemical/physiological study [60].

Most of the evidence for the function of ATPases, such as Cu-ATPases rests on indirect evidence, such as gene induction by copper, conferring of copper tolerance and/or loss of copper tolerance in a KO mutant. For only a few ATPases was transport demonstrated directly. CopB of *E. hirae* was the first Cu-ATPase for which transport was directly demonstrated, using $^{64}Cu^+$ in membrane vesicles [61]. It could be shown that Cu^+, rather than Cu^{2+}, was the substrate of the enzyme. Ag^+, which is a mimetic of Cu^+, was transported by CopB at the same rate and with the same affinity as Cu^+. Given the evolutionary conservation of copper ATPases from bacteria to man, it is likely that Cu^+ is the substrate of all copper ATPases.

One report of Cu^{2+} rather than Cu^+-transport by CopB of *Archaeoglobus fulgidus* has appeared [62], but could not be verified in this authors laboratory. *A. fulgidus* CopB is active at 75 °C, but at that temperature, Cu^{2+} greatly stimulates spontaneous ATP hydrolysis; so the observation of Cu^{2+}-transport is probably an artifact (M. Solioz, unpublished observations). Cu^{2+}-transport by an ATPase has also been reported for microsomes [63], but has never been verified. Voskoboinik et al. [64] demonstrated Cu^+ transport by the human Menkes Cu-ATPase (ATP7A) in membrane vesicles. This system later allowed the authors to analyzed the impact of Menkes disease mutations or mutations in the N-terminal MBDs on the function of the ATPase [65, 66]. Ag^+ also appears to be transported by the eukaryotic copper ATPases since Ag^+ shows a distribution similar to copper in tissues, the liver, and milk [67].

The reasons for the relatively few copper transport studies that have been conducted lies in technical handicaps. First, there are no convenient copper isotopes: ^{64}Cu is relatively easy to produce, but has a half-life of only 13 h, so its use is restricted to a few labs in the vicinity of a research reactor; ^{67}Cu with a more convenient half-life of 62 h is difficult to make and thus hard to obtain. $^{110m}Ag^+$ with a half-life of 250 d would be an extremely useful isotope for copper research, but is, for unknown reasons, no longer available anywhere in the Western world. As an alternative, some labs have turned to the use of the natural, stable isotopes, ^{63}Cu and ^{65}Cu, but this requires more expensive instrumentation and does not provide tracer-sensitivity. Secondly, the reconstitution of purified membrane proteins into artificial liposomes is a dying art, only still practiced by a few aficionados.

More recent work has address the *mechanism* of copper transport by Cu-ATPases. In addition to the N-terminal MBD, all copper ATPases feature a conserved CPC, CPN, or CPS motif in transmembrane helix 6. This motif has been shown to be part of two copper binding sites which are buried in the membrane. In CopA of *A. fulgidus*, C380, C382, and Y682 were postulated to form site 1, and D683, M711, and S715 site 2 and the transport of two copper ions per cycle has been postulated [58, 68]. The six amino acid residues for the two copper sites are distributed between transmembrane helices 6, 7, and 8. These findings were based on site-directed mutagenesis and modelling, but in absence of a three-dimensional structure of the ATPase. The binding of two Cu^+ per transport site was also measure for the two *Arabidopsis thaliana* copper ATPases in a static state [69].

A different concept was derived in more recent biochemical and molecular-dynamics study of *L. pneumophila* CopA, for which a structure is available [51]. In this ATPase, a copper 'entry' site was postulated to be formed by M148 and C382 and an intramembranous 'exit' site formed by C384, E689, and M717 (Fig. 3.10) [70]. A different architecture of the two copper sites was proposed by Mattle et al. [71], involving M148, C382 and C384 for the copper entry site and C382, C384 and M717 for the exit site. Recent measurements of the copper transport stoichiometry of *E. coli* CopA in giant unilamellar liposomes arrived at a transport stoichiometry of one Cu^+ per ATP [72]. Clearly, more work is required to understand copper transport by P1B-type ATPases at the molecular level.

The well characterized eukaryotic P-type ATPases all catalyze an antiport mechanism, namely 3 Na^+/2 K^+-, K^+/H^+-, and Ca^{2+}/2 H^+-exchange. Based on the current mechanistic principles of P-type ATPases, it could be expected that copper ATPases catalyze electroneutral Cu^+/H^+-exchange. The Cu^+ ion will require a negative charge at the binding site. Cu^+ release could then be thermodynamically facilitate by a proton entering that site. However, this will be difficult to show experimentally. For one, only few laboratories still master the art of membrane protein reconstitution and transport measurements with vesicles. In addition, copper ATPases have much lower turnover rates than ATPases transporting alkali metal ions, which make proton flux measurements very difficult because of the high H^+ leak rate of reconstituted membrane vesicles.

Cu-ATPases transport Cu^+ by the standard P-type ATPase mechanism [47, 48]. It is now clear that Cu^+ ions bound to the N-terminal ATPase MBDs are not for transport, but fulfil a poorly understood regulatory role (Fig. 3.10b). In vitro, purified copper ATPases can be stimulated by nano- to micromolar concentrations of copper [74]. In the cell, free Cu^+ may present at concentrations as low as 10^{-21} M [75] and Cu^+ for transport must be scavenged and delivered to the ATPase by copper chaperones. These will dock to a special 'platform' on the ATPase for Cu^+ delivery. Transfer of the Cu^+ to the ATPase involves protein-protein interaction and exploits a gradient of increasing copper-binding affinity [76]. Cu^+ is released from the chaperone via ligand exchange and transferred to the ATPase. Cu^+ transport is coupled to ATP hydrolysis, which provides the energy to pump against a concentration gradient.

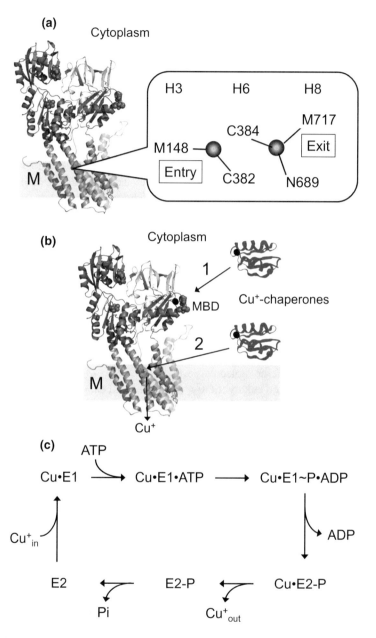

Fig. 3.10 Concept of copper pumping illustrated with CopA of *E. hirae*. A. Model of the CopA structure, with the position of the CPC motif in the membrane (M) indicated and a blow-up of the two intramembranous copper binding sites. The Entry site is formed by M148 and C382, and the Exit site by C384, N689, and M717; H6, H7, and H8 indicate the helices harboring these amino acids. B. The two different entry pathways for copper: 1, delivery of Cu$^+$ to the MBD in the N-terminus for regulatory purposes; 2, Cu$^+$ delivery to the Cu$^+$-transport site. C. Proposed reaction cycle for bacterial copper ATPases (Adapted from Ref. [73]). See text for explanations

During pumping, the enzyme cycles between two conformations, E1 and E2, involving a number of discrete steps (Fig. 3.10c). Cytoplasmic copper binds to the E2 form of the ATPase, inducing a conformational change to Cu•E1. Following ATP-binding to form Cu•E1•ATP, the γ-phosphate of ATP is transferred to the aspartic acid residue of the conserved DKTGT motif under conservation of the high-energy phosphate bond, resulting in Cu•E1~P•ADP. The formation of this acyl-phosphate intermediate, generally designated as '~P' (pronounced "squiggle-P") is a unique feature of P-type ATPases. Next, ADP is released from the enzyme and the energy of the high-energy acylphosphate bond on the enzyme is used to accomplish a conformational change to Cu•E2-P, from which Cu^+ is released to the outside of the cell, forming E2-P. Finally, Pi (inorganic phosphate) is released, taking the enzyme back to the E2 form.

3.5 Copper-Regulated Gene Expression

In contrast to copper ATPases, which are closely related across phyla, copper-responsive regulators of gene expression form a diverse group of proteins. As stated previously, once copper became a bioelement, copper-responsive regulators of gene expression had to evolve. This probably happened by convergent evolution of pre-existing regulatory protein, which would explain the range of structurally unrelated copper-responsive regulators we see today.

Currently, five families of structurally and functionally distinct copper-dependent transcriptional regulators are known [77]. They are named by their founding members ArsR-, MerR-, CsoR-, CopY-, or TetR-type regulators (Fig. 3.11). The CopY-type repressor will be discussed in detail below and gene regulation by CsoR- and CueR-type repressors and by two-component regulatory systems (CusRS) which primarily occur in Gram-negative bacteria will be described in Chap. 4.

CopY of E. hirae is the archetype of copper-responsive transcriptional regulators present in many Firmicutes [19, 78]. The NMR structure of the DNA-binding domain of CopY revealed that it is a winged-helix protein [79]. The native CopY was too insoluble for structural work, so only a structure of the N-terminal DNA binding domain could be obtained [79]. In the Zn(II)-form, CopY is dimeric and binds to the cop promoter, which features the consensus binding motif TACAnnTGTA. If copper becomes excessive in the cytoplasm, Cu(I)CopZ donates Cu^+ to the repressor. This displaces the bound Zn^{2+} in a ligand-transfer mechanism (Fig. 3.12) and converts the repressor to the $Cu(I)_2CopY$ form, which loses DNA binding affinity, monomerizes, and eventually gets degraded (cf. Fig. 3.2) [80–82]. This allows transcription of the cop operon to proceed. X-ray absorption studies showed that the Cu^+ ions moved from, what was best fit by a diagonal coordination in CopZ, to an unambiguously trigonal coordination in CopY, in line with the expected formation of a compact $(Cu^+)_2$-thiolate in the CXCXXXXCXC metal coordination site of CopY [82].

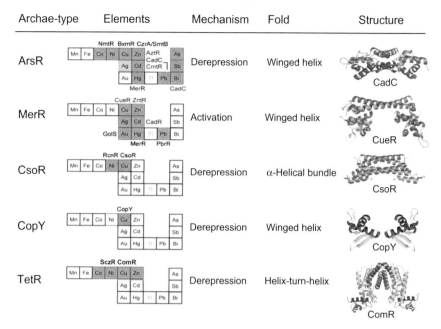

Fig. 3.11 Overview of copper-responsive activators and repressors of gene expression. The archaea-types are named after their founding members, which were sometimes not copper-responsive regulators. The column *Elements* shows the transition metals for which a repressor of that particular archaea-type is known. MerR is the only regulator of the group which is not released from the DNA by copper, but undergoes a conformational change that activates transcription. Shown are also cartoons of the structures and other common names

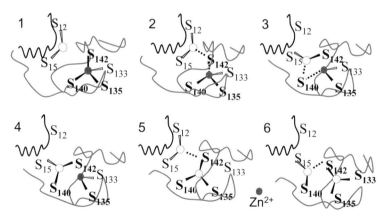

Fig. 3.12 Ligand transfer mechanism in the activation of CopY by Cu(I)CopZ. 1–6, sequential steps in the transfer of two Cu$^+$ to replace the single Zn^{2+} of CopY. S12 and S15 are the cysteine residues of CopZ (black lines) which bind Cu$^+$, and S133, S135, S140, and S142 are the cysteine residues of CopY (purple lines) which bind one Zn^{2+} in the DNA-binding form or two Cu$^+$ ions for release from the DNA. Cu$^+$ is represented by yellow spheres and Zn^{2+} by red spheres

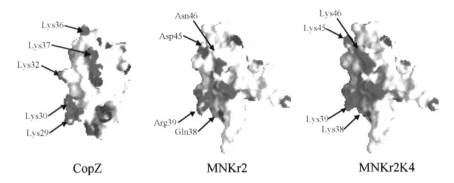

Fig. 3.13 Images of the surface charges of CopZ, MNKr2, and MNKr2K4. Positively charged surfaces are blue, negatively charged ones are red. The lysine residues of CopZ presumed to be critical for interaction with CopY are indicated. On MNKr2, the amino acid residues which were mutated to lysine are shown. MNKr2K4 shows the gain-of-function mutant with the lysine residues introduced by site-directed mutagenesis

By surface plasmon resonance analysis, it could be shown that copper transfer from CopZ to CopY involved protein-protein interaction and affinity and rate constants could be derived for individual interaction steps [83]. Structural work on the yeast chaperone, ATX1, and its cognate target, the N-terminal domain of the CCC2 Cu-ATPase, had yielded a detailed picture of the interaction of the two protein [42]. The interaction was found to be mainly of an electrostatic nature, with some hydrogen bonds stabilizing the complex. However, CopZ exhibits a different pattern of charged residues. To mimic CopZ function with another chaperone, MNKr2, the second CopZ-like HMB domain of the human Menkes Cu-ATPase (amino acids 164–181 of ATP7A), was mutated such that four lysine residues were in positions similar to the four lysines of CopZ (mutations Q38K, R39K, D45K, and N46K, Fig. 3.13). This mutant MNKr2K4 could actually activate the CopY repressor, just like CopZ, by donating Cu^+ to CopY [82]. This gain-of-function mutation experiment highlights two important concepts: first, protein-protein interaction defines the target of a chaperone, and secondly, MBDs in the N-termini of Cu-ATPases are indeed chaperone equivalents.

3.6 Copper Buffering/Storage

GSH or alternative low-molecular weight thiols like mycothiol, bacillithiol, or γ-L-glutamyl-L-cysteine have been identified in the cytoplasm of most prokaryotes [84–87]. GSH and other thiols are in equilibrium with the oxidized, dimeric form, which is GSSG in the case of GSH. The GSH/GSSG redox couple helps to maintain a negative cytoplasmic redox potential in growing cells [88]. In the absence of GSH, another small-molecular weight thiol in all likelihood takes the role of GSH.

It was shown that under anaerobic conditions, GSH enhances copper toxicity through the formation of GS–Cu–SG complexes (two GSH molecules complexing one Cu) which presumably donate copper to enzyme sites reserved for other transition metals (see Chap. 2) [89]. However, under aerobic conditions, complexation of copper (and other heavy metal ions) appears to offer protection against toxicity (cf. Fig. 3.2) [90–93]. Why GS–Cu–SG complexes are toxic under anaerobic conditions, but detoxifying under aerobic conditions remains to be explore. It was shown that *E. coli* cells devoid of the GSH biosynthetic pathway were considerably more sensitive to killing by copper compared to wild-type cells [94]. The effect became particularly apparent in the absence of a functioning copper ATPase. *Streptococcus pneumoniae* cannot synthesize GSH, but can take it up via an ABC-type transporter [95]. When GSH was supplied to growing *S. pneumonia*, the cells became considerably more resistant to copper, but also to the non-redox-active heavy metals cadmium and zinc. GSH was also shown to increase resistance towards Hg^{2+} and arsenite (AsO^{2-}) in *E. coli* [90]. Clearly, GSH can prevent cells from heavy metal toxicity, especially if they are compromised in other ways, such as by mutation or other stresses conditions. But it must be emphasized that wild-type cells are quite well endowed to deal with various 'every-day' stress conditions and that GSH and similar systems should be viewed as back-up systems under special conditions.

In eukaryotic cells, metallothioneins play an important role in the defense against metal ion toxicity and can be considered a first-line defense [96]. Although metal-lothioneins are not routinely detected in Gram-positive bacteria, they do occur in some organisms. In *Mycobacterium tuberculosis*, a 4.9 kDa cysteine-rich protein which functions as a metallothionein was identified and termed MymT (mycobacterial metallothionein) [97]. The protein is induced by copper or cadmium and helps to protect the organism against copper toxicity. In vitro, MymT binds up to six copper ions in a solvent shielded environment, which is typical for metallothioneins. The amino acid sequence of MymT was found to be highly conserved among pathogenic mycobacteria.

Another cytosolic copper storage protein, Csp3, first discovered in *Methylosinus trichosporium* OB3b (see also Chap. 4), was found to also be present in *Bacillus subtilis* [98]. The crystal structures of *B. subtilis* and *M. trichosporium* Csp3 have been solved and revealed a remarkable structure of four 4-helix-bundle monomers, arranged in a tetrameric structure which can bind up to 80 copper atoms via thiolate bonds. Based on the absence of a translocation signal, it was assumed that the protein has a cytoplasmic location. Other bacterial copper storage proteins will surely still be discovered, but contrary to eukaryotic metallothioneins, they promise to be more diverse in bacteria than in eukaryotes [99].

3.7   Copper Loading of Cuproenzymes

Little is known about the metallation of cuproenzymes in Gram-positive organisms, but the topic is covered extensively in Chap. 4 for Gram-negative bacteria. The main reason for the limited knowledge of cuproenzyme formation in Gram-positives may lay in the fact that copper-containing proteins are not very widespread in these organisms. From the study of Gram-negative organisms, it has become clear that copper containing enzymes are either periplasmic or integral cytoplasmic membrane proteins. For the metallation of these proteins, copper is usually exported from the cytoplasm by dedicated 'assembly transporters', rather than drawn from the periplasmic copper pool. *E. hirae* possesses two copper ATPases, CopA and CopB, which are co-regulated by copper. [1, 2]. It was found that only CopB functions as a copper export ATPase that provides copper tolerance to the organism [57, 61]. It was later found that it is rather common for bacteria to have two or more copper ATPases which serve different functions: a high-turnover copper ATPase serves in copper extrusion to provide copper tolerance, while a second, low-turnover copper ATPase is required for metallation of one or several cuproenzymes in the periplasm or the cytoplasmic membrane (see Chap. 4). So in the case of *E. hirae*, CopB serves in copper tolerance, while CopA could be involved in the metallation of a cuproenzyme in the cytoplasmic membrane or the periplasm. However, no such enzyme has been discoverd to date and future work will have to show if this concept also holds true.

3.8 Copper Regulons and Accessory Defense Mechanisms

The CopR repressor of *L. lactis* is a CopY-type repressor (see Sect. 3.5) and like most of these, recognize a TACAnnTGTA inverted repeat, also called the '*cop*-box' (Fig. 3.14a) [100]. By performing a genome-wide search for *cop*-boxes in *L. lactis* IL1403, Magnani et al. [9] identified 14 genes, organized into four operons and two monocistronic genes, which are all under the control of CopR and thus copper-inducible. This so-called 'CopR Regulon' encompasses the *copRZA* copper resistance operon, which encodes the CopR repressor, the CopZ copper chaperone, and the CopA copper export ATPase (Fig. 3.14b). A monocistronic gene encodes a second copper ATPase, CopB. This enzyme could not be shown to confer copper tolerance to *L. lactis* and probably serves in the metallation of enzymes (see Sect. 3.7.).

Other genes of the CopR regulon have no direct association with copper. The function of several of these genes was elucidated and revealed fascinating 'accessory' copper resistance mechanisms. A putative function could also be assigned to the *lctO* gene, encoding a NAD-independent lactate oxidase. This enzyme catalyzes the reaction L-lactate $+ O_2 \rightarrow$ pyruvate $+ H_2O_2$. L-lactate is a waste-product of fermentative growth of *L. lactis* and is produced in large amounts. Lactate oxidation could serve to eliminate molecular oxygen under copper stress conditions and thus lower overall stress (Fig. 3.15) [101]. A similar mechanism has been proposed for the oxygen-consuming NADH oxidase of *Lactobacillus delbrueckii* subsp. *bulgaricus* [102].

The *ytjD* gene of the CopR regulon encodes a nitroreductase, which was named CinD (copper-induced nitroreductase D). CinD is a flavoprotein which can reduce

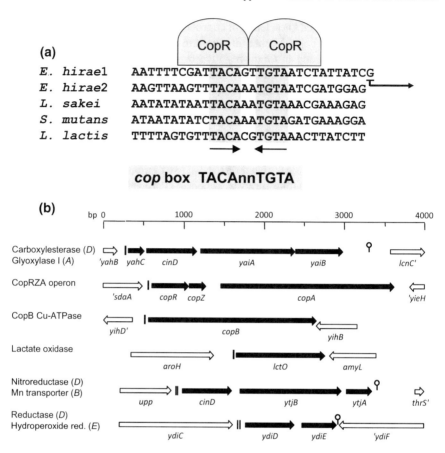

Fig. 3.14 **a** A dimer of the *L. lactis* CopR repressor recognizes the inverted repeat sequence TACAnnTGTA, also called '*cop* box'. CopR also binds to cop boxes from other Firmicutes, as indicated in the Figure. **b** Genes and operons (black arrows) constituting the CopR regulon. Open arrows are genes that do not belong to the CopR regulon and are not copper-regulated. The vertical lines indicate *cop* boxes, which can be present in one or two copies, and the lariats signify terminators. The genes are drawn to the scale indicated above the Figure

2,6-dichlorophenolindophenol, 4-nitroquinoline-N-oxide, and other nitroaromatic compounds, using NADH as reductant [103]. CinD is thus a copper-induced nitrore-ductase which can protect *L. lactis* against nitrosative stress that could be exerted by nitroaromatic compounds in the presence of copper.

Finally, an exciting function could also be assigned to some genes of the *yahCD-yaiAB* operon of the CopR regulon [104]. Hydroquinones occur naturally, but are also widely used in industrial activities. Hydroquinones alone are benign to bacteria, but copper catalyzes their rapid conversion to highly toxic *p*-benzoquinones. Combined quinone/copper exposure is thus a threat to bacteria that is alleviated by the copper-induced *yahCD-yaiAB* operon. It encodes YaiB, a flavoprotein that converts toxic *p*-

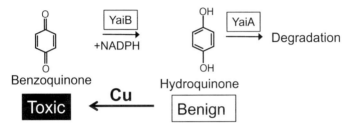

A. Reduction of oxidative stress

L-Lactate →(LctO, +O$_2$) Pyruvate

B. Reduction of nitrosative stress

2,6-Dichlorophenolindophenol → (+ 2H$^+$ + 2e$^-$, CinD, +NADH) Reduced DCPIP

C. Reduction of copper-hydroquinone stress

Benzoquinone → (YaiB, +NADPH) Hydroquinone → (YaiA) Degradation

Toxic ←(Cu)— Benign

Fig. 3.15 Accessory copper defense mechanisms. **a** LctO oxidase converts L-lactate to pyruvate, thereby consuming O$_2$, which could lead to ROS production in the presence of copper. **b** CinD reduces nitroaromatic compounds which can cause nitrosative stress in combination with copper. **c** Copper converts benign hydroquinones to toxic benzoquinones. YaiB reduces these back to hydroquinone and YaiA degrades these further. See text for additional details

benzoquinones back to benign hydroquinones, using NADPH as reductant, and YaiA, which degrades hydroquinones. Deletion of *yaiB* causes complete growth arrest of *L. lactis* under combined quinone/copper stress, showing the physiological importance of this operon. Importantly, this mechanism was revealed by a mini-systems biology approach and would not have been discovered by studies under laboratory growth conditions. Most lactococcal and many streptococcal species possess *yaiA* and *yaiB*-like genes, possibly due to their preferential association with plants, which are rich in quinones. Also, CopR-type regulons are widespread in bacteria, particularly in lactic acid bacteria, and more fascinating work can be awaited in this area.

A similar quinone resistance network was identified in *Bacillus subtilis*, but rather than by copper, it is activated via redox sensing by two repressors, YodB and CatR [105]. YodB controls the expression azoreductase AzoR1, the nitroreductase YodC, and the Spx regulator. In addition, YodB and CatR together also control the expression of the oxidoreductase CatD, and the thiol-dependent dioxygenase CatE. Another repressor, MhqR, controls the expression of a set of paralogous genes, but the mechanism of induction remains unknown. Clearly, there is a wealth of accessory mechanisms to control copper toxicity in bacteria. Such mechanisms might often not be relevant to standard laboratory growth conditions and are therefore difficult to discover, but they are most likely decisive for survival in natural habitats and/or in complex bacterial communities.

References

1. Odermatt A, Suter H, Krapf R et al (1992) An ATPase operon involved in copper resistance by *Enterococcus hirae*. Ann NY Acad Sci 671:484–486
2. Odermatt A, Suter H, Krapf R et al (1993) Primary structure of two P-type ATPases involved in copper homeostasis in *Enterococcus hirae*. J Biol Chem 268:12775–12779
3. Itoh T, Takemoto K, Mori H et al (1999) Evolutionary instability of operon structures disclosed by sequence comparisons of complete microbial genomes. Mol Biol Evol 16:332–346
4. Shi L, Dong H, Reguera G et al (2016) Extracellular electron transfer mechanisms between microorganisms and minerals. Nat Rev Microbiol 14:651–662
5. Abicht HK, Gonskikh Y, Gerber SD et al (2013) Non-enzymatic copper reduction by menaquinone enhances copper toxicity in *Lactococcus lactis* IL1403. Microbiology 159:1190–1197
6. Tachon S, Brandsma JB, Yvon M (2010) NoxE NADH oxidase and the electron transport chain are responsible for the ability of *Lactococcus lactis* to decrease the redox potential of milk. Appl Environ Microbiol 76:1311–1319
7. Duwat P, Sourice S, Cesselin B et al (2001) Respiration capacity of the fermenting bacterium *Lactococcus lactis* and its positive effects on growth and survival. J Bacteriol 183:4509–4516
8. Solioz M, Vulpe C (1996) CPx-type ATPases: a class of P-type ATPases that pump heavy metals. Trends Biochem Sci 21:237–241
9. Magnani D, Barré O, Gerber SD et al (2008) Characterization of the CopR regulon of *Lactococcus lactis* IL1403. J Bacteriol 190:536–545
10. Tottey S, Rich PR, Rondet SA et al (2001) Two Menkes-type atpases supply copper for photosynthesis in Synechocystis PCC 6803. J Biol Chem 276:19999–20004
11. Phung LT, Ajlani G, Haselkorn R (1994) P-type ATPase from the cyanobacterium *Synechococcus* 7942 related to the human Menkes and Wilson disease gene products. Proc Natl Acad Sci USA 91:9651–9654
12. Lewinson O, Lee AT, Rees DC (2009) A P-type ATPase importer that discriminates between essential and toxic transition metals. Proc Natl Acad Sci USA 106:4677–4682
13. Chillappagari S, Miethke M, Trip H et al (2009) Copper acquisition is mediated by YcnJ and regulated by YcnK and CsoR in *Bacillus subtilis*. J Bacteriol 191:2362–2370
14. Fukuhara T, Kobayashi K, Kanayama Y et al (2016) Identification and characterization of the *zosA* gene involved in copper uptake in *Bacillus subtilis* 168. Biosci Biotechnol Biochem 80:600–609
15. Hirooka K, Edahiro T, Kimura K et al (2012) Direct and indirect regulation of the *ycnKJI* operon involved in copper uptake through two transcriptional repressors, YcnK and CsoR, in *Bacillus subtilis*. J Bacteriol 194:5675–5687
16. Wang L, Zhu M, Zhang Q et al (2017) Diisonitrile natural product SF2768 functions as a chalkophore that mediates copper acquisition in *Streptomyces thioluteus*. ACS Chem Biol 12:3067–3075
17. Capdevila DA, Edmonds KA, Giedroc DP (2017) Metallochaperones and metalloregulation in bacteria. Essays Biochem 61:177–200
18. Lin SJ, Culotta VC (1995) The ATX1 gene of *Saccharomyces cerevisiae* encodes a small metal homeostasis factor that protects cells against reactive oxygen toxicity. Proc Natl Acad Sci USA 92:3784–3788
19. Odermatt A, Solioz M (1995) Two *trans*-acting metalloregulatory proteins controlling expression of the copper-ATPases of *Enterococcus hirae*. J Biol Chem 270:4349–4354

20. Barkay T, Miller SM, Summers AO (2003) Bacterial mercury resistance from atoms to ecosystems. FEMS Microbiol Rev 27:355–384

21. Morby AP, Hobman JL, Brown NL (1995) The role of cysteine residues in the transport of mercuric ions by the Tn501 MerT and MerP mercury-resistance proteins. Mol Microbiol 17:25–35

22. Fan B, Grass G, Rensing C et al (2001) *Escherichia coli* CopA N-terminal Cys(X)$_2$Cys motifs are not required for copper resistance or transport. Biochem Biophys Res Commun 286:414–418

23. Hou ZZ, Narindrasorasak S, Bhushan B et al (2001) Functional analysis of chimeric proteins of the Wilson Cu(I)-ATPase (ATP7B) and ZntA, a Pb(II)/Zn(II)/Cd(II)-ATPase from *Escherichia coli*. J Biol Chem e-pub

24. Mattle D, Sitsel O, Autzen HE et al (2013) On allosteric modulation of P-type Cu-ATPases. J Mol Biol 425:229–2308

25. Wimmer R, Herrmann T, Solioz M et al (1999) NMR structure and metal interactions of the CopZ copper chaperone. J Biol Chem 274:22597–22603

26. Banci L, Bertini I, Cantini F et al (2009) An NMR study of the interaction of the N-terminal cytoplasmic tail of the Wilson disease protein with copper(I)-HAH1. J Biol Chem 284:9354–9360

27. Singleton C, Banci L, Ciofi-Baffoni S et al (2008) Structure and Cu(I)-binding properties of the N-terminal soluble domains of *Bacillus subtilis* CopA. Biochem J 411:571–579

28. Banci L, Bertini I, Chasapis CT et al (2007) Interaction of the two soluble metal-binding domains of yeast Ccc2 with copper(I)-Atx1. Biochem Biophys Res Commun 364:645–649

29. Achila D, Banci L, Bertini I et al (2006) Structure of human Wilson protein domains 5 and 6 and their interplay with domain 4 and the copper chaperone HAH1 in copper uptake. Proc Natl Acad Sci USA 103:5729–5734

30. Banci L, Bertini I, Cantini F et al (2005) A NMR study of the interaction of a three-domain construct of ATP7A with copper(I) and copper(I)-HAH1: the interplay of domains. J Biol Chem 280:38259–38263

31. Banci L, Bertini I, Ciofi-Baffoni S et al (2005) An NMR study of the interaction between the human copper(I) chaperone and the second and fifth metal-binding domains of the Menkes protein. FEBS J 272:865–871

32. Banci L, Bertini I, Ciofi-Baffoni S et al (2004) Solution structures of a cyanobacterial metallochaperone: insight into an atypical copper-binding motif. J Biol Chem 279:27502–27510

33. Anastassopoulou I, Banci L, Bertini I et al (2004) Solution structure of the apo and copper(I)-loaded human metallochaperone HAH1. Biochemistry 43:13046–13053

34. Banci L, Bertini I, Del Conte R et al (2003) X-ray absorption and NMR spectroscopic studies of CopZ, a copper chaperone in *Bacillus subtilis*: the coordination properties of the copper ion. Biochemistry 42:2467–2474

35. Banci L, Bertini I, Del Conte R et al (2001) Copper trafficking: the solution structure of *Bacillus subtilis* CopZ. Biochemistry 40:15660–15668

36. Arnesano F, Banci L, Bertini I et al (2001) Solution structure of the Cu(I) and apo forms of the yeast metallochaperone, Atx1. Biochemistry 40:1528–1539

37. Jordan IK, Natale DA, Koonin EV et al (2001) Independent evolution of heavy metal-associated domains in copper chaperones and copper-transporting ATPases. J Mol Evol 53:622–633

38. Rosenzweig AC (2001) Copper delivery by metallochaperone proteins. Acc Chem Res 34:119–128

39. Boal AK, Rosenzweig AC (2009) Structural biology of copper trafficking. Chem Rev 109:4760–4779

40. Kay KL, Zhou L, Tenori L et al (2017) Kinetic analysis of copper transfer from a chaperone to its target protein mediated by complex formation. Chem Commun (Camb) 53:1397–1400

41. Banci L, Bertini I, Ciofi-Baffoni S et al (2006) The delivery of copper for thylakoid import observed by NMR. Proc Natl Acad Sci USA 103:8320–8325

42. Arnesano F, Banci L, Bertini I et al (2004) A docking approach to the study of copper trafficking proteins; interaction between metallochaperones and soluble domains of copper ATPases. Structure (Camb) 12:669–676
43. Banci L, Bertini I, Ciofi-Baffoni S et al (2003) Understanding copper trafficking in bacteria: interaction between the copper transport protein CopZ and the N-terminal domain of the copper ATPase CopA from *Bacillus subtilis*. Biochem 42:1939–1949
44. Lutsenko S (2016) Copper trafficking to the secretory pathway. Metallomics 8:840–852
45. Harrison MD, Jones CE, Solioz M et al (2000) Intracellular copper routing: the role of copper chaperones. Trends Biochem Sci 25:29–32
46. Robinson NJ, Winge DR (2010) Copper metallochaperones. Annu Rev Biochem 79:537–562
47. Pedersen PL, Carafoli E (1987) Ion motive ATPases. I. Ubiquity, properties, and significance to cell function. Trends Biochem Sci 12:146–150
48. Pedersen PL, Carafoli E (1987) Ion motive ATPases. II. Energy coupling and work output. Trends Biochem Sci 12:186–189
49. Toyoshima C, Nakasako M, Nomura H et al (2000) Crystal structure of the calcium pump of sarcoplasmic reticulum at 2.6 Å resolution. Nature 405:647–655
50. Lutsenko S, Kaplan JH (1995) Organization of P-type ATPases: significance of structural diversity. Biochemistry 34:15607–15613
51. Gourdon P, Liu XY, Skjorringe T et al (2011) Crystal structure of a copper-transporting PIB-type ATPase. Nature 475:59–64
52. Lübben M, Portmann R, Kock G et al (2009) Structural model of the CopA copper ATPase of *Enterococcus hirae* based on chemical cross-linking. Biometals 22:363–375
53. Smith AT, Smith KP, Rosenzweig AC (2014) Diversity of the metal-transporting P-type ATPases. J Biol Inorg Chem 19:947–960
54. Takatsu H, Tanaka G, Segawa K et al (2014) Phospholipid flippase activities and substrate specificities of human type IV P-type ATPases localized to the plasma membrane. J Biol Chem 289:33543–33556
55. Axelsen KB, Palmgren MG (1998) Evolution of substrate specificities in the P-type ATPase superfamily. J Mol Evol 46:84–101
56. Ogawa H, Haga T, Toyoshima C (2000) Soluble P-type ATPase from an archaeon, *Methanococcus jannaschii*. FEBS Lett 471:99–102
57. Odermatt A, Krapf R, Solioz M (1994) Induction of the putative copper ATPases, CopA and CopB, of *Enterococcus hirae* by Ag^+ and Cu^{2+}, and Ag^+ extrusion by CopB. Biochem Biophys Res Commun 202:44–48
58. Raimunda D, Gonzalez-Guerrero M, Leeber BW III et al (2011) The transport mechanism of bacterial Cu^+-ATPases: distinct efflux rates adapted to different function. Biometals 24:467–475
59. Raimunda D, Padilla-Benavides T, Vogt S et al (2013) Periplasmic response upon disruption of transmembrane Cu transport in *Pseudomonas aeruginosa*. Metallomics 5:144–151
60. Bork P (2000) Powers and pitfalls in genome analysis: the 70% hurdle. Genome Res 10:398–400
61. Solioz M, Odermatt A (1995) Copper and silver transport by CopB-ATPase in membrane vesicles of *Enterococcus hirae*. J Biol Chem 270:9217–9221
62. Mana-Capelli S, Mandal AK, Arguello JM (2003) *Archaeoglobus fulgidus* CopB is a thermophilic Cu^{2+}-ATPase: functional role of its histidine-rich-N-terminal metal binding domain. J Biol Chem 278:40534–40541
63. Bingham MJ, Ong TJ, Ingledew WJ et al (1996) ATP-dependent copper transporter, in the golgi apparatus of rat hepatocytes, transports Cu(II) not Cu(I). Am J Physiol 271:G741–G746
64. Voskoboinik I, Brooks H, Smith S et al (1998) ATP-dependent copper transport by the Menkes protein in membrane vesicles isolated from cultured Chinese hamster ovary cells. FEBS Lett 435:178–182
65. Voskoboinik I, Mar J, Camakaris J (2003) Mutational analysis of the Menkes copper P-type ATPase (ATP7A). Biochem Biophys Res Commun 301:488–494

66. Voskoboinik I, Strausak D, Greenough M et al (1999) Functional analysis of the N-terminal CXXC metal-binding motifs in the human menkes copper-transporting P-type ATPase expressed in cultured mammalian cells. J Biol Chem 274:22008–22012

67. Hanson SR, Donley SA, Linder MC (2001) Transport of silver in virgin and lactating rats and relation to copper. J Trace Elem Med Biol 15:243–253

68. Gonzalez-Guerrero M, Eren E, Rawat S et al (2008) Structure of the two transmembrane Cu^+ transport sites of the Cu^+-ATPases. J Biol Chem 283:29753–29759

69. Blaby-Haas CE, Padilla-Benavides T, Stube R et al (2014) Evolution of a plant-specific copper chaperone family for chloroplast copper homeostasis. Proc Natl Acad Sci USA 111:E5480–E5487

70. Grønberg C, Sitsel O, Lindahl E et al (2016) Membrane anchoring and ion-entry dynamics in P-type ATPase copper transport. Biophys J 111:2417–2429

71. Mattle D, Zhang L, Sitsel O et al (2015) A sulfur-based transport pathway in Cu^+-ATPases. EMBO Rep 16:728–740

72. Wijekoon CJ, Udagedara SR, Knorr RL et al (2017) Copper ATPase CopA from *E. coli*. Quantitative correlation between ATPase activity and vectorial copper transport. J Am Chem Soc 139:4266–4269

73. Hatori Y, Lewis D, Toyoshima C et al (2009) Reaction cycle of *Thermotoga maritima* copper ATPase and conformational characterization of catalytically deficient mutants. Biochemistry 48:4871–4880

74. Bissig K-D, Voegelin TC, Solioz M (2001) Tetrathiomolybdate inhibition of the *Enterococcus hirae* CopB copper ATPase. FEBS Lett 507:367–370

75. Changela A, Chen K, Xue Y et al (2003) Molecular basis of metal-ion selectivity and zepto-molar sensitivity by CueR. Science 301:1383–1387

76. Banci L, Bertini I, Ciofi-Baffoni S et al (2010) Affinity gradients drive copper to cellular destinations. Nature 465:645–650

77. Giedroc DP, Arunkumar AI (2007) Metal sensor proteins: nature's metalloregulated allosteric switches. Dalton Trans 3107–3120

78. Strausak D, Solioz M (1997) CopY is a copper-inducible repressor of the *Enterococcus hirae* copper ATPases. J Biol Chem 272:8932–8936

79. Cantini F, Banci L, Solioz M (2009) The copper-responsive repressor CopR of *Lactococcus lactis* is a 'winged helix' protein. Biochem J 417:493–499

80. Cobine P, Wickramasinghe WA, Harrison MD et al (1999) The *Enterococcus hirae* copper chaperone CopZ delivers copper(I) to the CopY repressor. FEBS Lett 445:27–30

81. Lu ZH, Solioz M (2001) Copper-induced proteolysis of the CopZ copper chaperone of *Enterococcus hirae*. J Biol Chem 276:47822–47827

82. Cobine PA, George GN, Jones CE et al (2002) Copper transfer from the Cu(I) chaperone, CopZ, to the repressor, Zn(II)CopY: metal coordination environments and protein interactions. Biochemistry 41:5822–5829

83. Portmann R, Magnani D, Stoyanov JV et al (2004) Interaction kinetics of the copper-responsive CopY repressor with the *cop* promoter of *Enterococcus hirae*. J Biol Inorg Chem 9:396–402

84. Fahey RC, Brown WC, Adams WB et al (1978) Occurrence of glutathione in bacteria. J Bacteriol 133:1126–1129

85. Newton GL, Arnold K, Price MS et al (1996) Distribution of thiols in microorganisms: mycothiol is a major thiol in most actinomycetes. J Bacteriol 178:1990–1995

86. Gaballa A, Newton GL, Antelmann H et al (2010) Biosynthesis and functions of bacillithiol, a major low-molecular-weight thiol in Bacilli. Proc Natl Acad Sci USA 107:6482–6486

87. Kim EK, Cha CJ, Cho YJ et al (2008) Synthesis of γ-glutamylcysteine as a major low-molecular-weight thiol in lactic acid bacteria *Leuconostoc* spp. Biochem Biophys Res Commun 369:1047–1051

88. Schafer FQ, Buettner GR (2001) Redox environment of the cell as viewed through the redox state of the glutathione disulfide/glutathione couple. Free Radic Biol Med 30:1191–1212

89. Obeid MH, Oertel J, Solioz M et al (2016) Mechanism of attenuation of uranyl toxicity by glutathione in *Lactococcus lactis*. Appl Environ Microbiol 82:3563–3571
90. Latinwo LM, Donald C, Ikediobi C et al (1998) Effects of intracellular glutathione on sensitivity of *Escherichia coli* to mercury and arsenite. Biochem Biophys Res Commun 242:67–70
91. Fu RY, Bongers RS, van Swam II et al (2006) Introducing glutathione biosynthetic capability into *Lactococcus lactis* subsp. *cremoris* NZ9000 improves the oxidative-stress resistance of the host. Metab Eng 8:662–671
92. Li Y, Hugenholtz J, Abee T et al (2003) Glutathione protects *Lactococcus lactis* against oxidative stress. Appl Environ Microbiol 69:5739–5745
93. Zhang J, Fu RY, Hugenholtz J et al (2007) Glutathione protects *Lactococcus lactis* against acid stress. Appl Environ Microbiol 73:5268–5275
94. Helbig K, Bleuel C, Krauss GJ et al (2008) Glutathione and transition-metal homeostasis in *Escherichia coli*. J Bacteriol 190:5431–5438
95. Potter AJ, Trappetti C, Paton JC (2012) *Streptococcus pneumoniae* uses glutathione to defend against oxidative stress and metal ion toxicity. J Bacteriol 194:6248–6254
96. Vasak M, Meloni G (2011) Chemistry and biology of mammalian metallothioneins. J Biol Inorg Chem 16:1067–1078
97. Gold B, Deng H, Bryk R et al (2008) Identification of a copper-binding metallothionein in pathogenic mycobacteria. Nat Chem Biol 4:609–616
98. Vita N, Landolfi G, Basle A et al (2016) Bacterial cytosolic proteins with a high capacity for Cu(I) that protect against copper toxicity. Sci Rep 6:39065
99. Blindauer CA (2011) Bacterial metallothioneins: past, present, and questions for the future. J Biol Inorg Chem 16:1011–1024
100. Portmann R, Poulsen KR, Wimmer R et al (2006) CopY-like copper inducible repressors are putative 'winged helix' proteins. Biometals 19:61–70
101. Barré O, Mourlane F, Solioz M (2007) Copper induction of lactate oxidase of *Lactococcus lactis*: a novel metal stress response. J Bacteriol 189:5947–5954
102. Marty-Teysset C, de la Torre F, Garel J (2000) Increased production of hydrogen peroxide by *Lactobacillus delbrueckii* subsp. *bulgaricus* upon aeration: involvement of an NADH oxidase in oxidative stress. Appl Environ Microbiol 66:262–267
103. Mermod M, Mourlane F, Waltersperger S et al (2010) Structure and function of CinD (YtjD) of *Lactococcus lactis*, a copper-induced nitroreductase involved in defense against oxidative stress. J Bacteriol 192:4172–4180
104. Mancini S, Abicht HK, Gonskikh Y et al (2015) A copper-induced quinone degradation pathway provides protection against combined copper/quinone stress in *Lactococcus lactis* IL1403. Mol Microbiol 95:645–659
105. Leelakriangsak M, Huyen NT, Towe S et al (2008) Regulation of quinone detoxification by the thiol stress sensing DUF24/MarR-like repressor, YodB in *Bacillus subtilis*. Mol Microbiol 67:1108–1124

Chapter 4
Copper Homeostasis in Gram-Negative Bacteria

Abstract Copper homeostasis in Gram-negative organisms is complicated by the two cell membranes and the periplasmic space. The thinking of how copper enters bacteria like *E. coli* has undergone some changes recently and new concepts have emerged. Also, the elusive CopZ-like copper chaperone has finally been discovered in *E. coli*. While the extrusion of excess copper via the CopA copper ATPase and the CusCFBA transporter and the regulation of these systems appear fairly clear, there are still major open questions concerning the metallation of cuproenzymes. Some provocative new concepts will be proposed.

Keywords Gram-negative · *Escherichia coli* · Two-component regulator RND-transporter · Periplasm · Multicopper oxidase · Copper uptake Photosynthesis · Cyanobacteria · Methanogens

Gram-negative copper homeostasis is best understood in *E. coli*, which is here used as the paradigm; other organisms are discussed to highlight special aspects and differences. There are still no general concepts for copper uptake by Gram-positive bacteria. Some specific uptake systems that have been reported will be discussed. Very recent work for the first time describes copper uptake by *E. coli* with a chalkophore, a mechanism previously only known for methanogens (Fig. 4.1). Such a mechanism could poof to be more universal than anticipated. Copper entering the cytoplasm is complexed by the CopZ-like copper chaperone, which directs it to regulators of gene expression and the CopA ATPases for export into the periplasmic space. An RND-type transporters, CusCFBA, extrudes periplasmic copper to the outside of the cell with the help of a special periplasmic copper chaperone, CusF. Additional defense against copper is provided by the periplasmic CueO-type multicopper oxidases, which can oxidized Cu^+ to less toxic Cu^{2+} and catechols to copper-binding pigments. Finally, the ComR/ComC system can reduce the permeability of the outer membrane for copper. Three different type of copper-responsive regulatory systems are involved in this network: a CsoR- and a TetR-type (ComR) regulator, and the CusRS two-component regulatory system. The metallation of cuproenzymes will also be discussed, with special emphasis on photosynthetic cyanobacteria and methanogens.

© The Author(s) 2018   49
M. Solioz, *Copper and Bacteria*, SpringerBriefs in Biometals,
https://doi.org/10.1007/978-3-319-94439-5_4

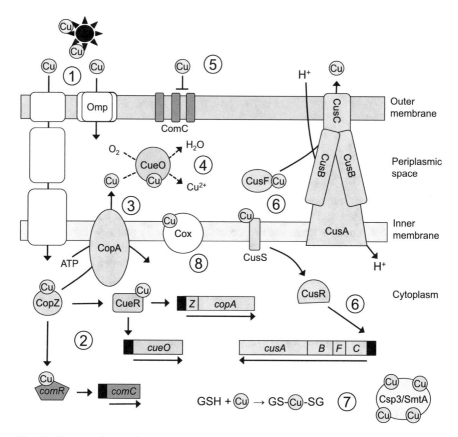

Fig. 4.1 Cartoon of copper homeostasis in *E. coli*. 1. Copper can traverse the outer membrane non-specifically *via* porins, and the plasma membrane fortuitously as a complex with organic substrates; 2. cytoplasmic copper is complexed by the CopZ copper chaperone and induces the expression of the CopA copper efflux ATPase, the CueO multicopper oxidase, and the ComC outer membrane protein; 3. CopA pumps excess cytoplasmic copper to the periplasmic space; 4. Cu^+ gets oxidized to less toxic Cu^{2+}; 5. ComC reduces nonspecific copper entry into the periplasm; 6. excess periplasmic copper induces, *via* CusRS, the $CusC_3B_6A_3$ copper transporter and the CusF copper chaperone, which together transport periplasmic copper across the outer membrane; 7. GSH and copper binding proteins interact with copper for detoxification or safe storage; 8. Cytochrome *c* oxidase, Cox, located in the cytoplasmic membrane, is metallated from the periplasm

4.1 Copper Uptake by Gram-Negative Bacteria

4.1.1 ATPase-Driven Copper Uptake

There have been a number of claims of copper acquisition by P1B-type copper ATPases (e.g. [1–6]). These and other reports on ATPase-catalyzed copper uptake were based on indirect observations, never direct copper transport, and were single reports without any follow-up work. It is probably safe to conclude that they were misinterpretations of complex phenomena. A large body of more recent evidence supports the view that all P1B-type copper ATPases catalyze copper export.

4.1.2 *Chalkophore Mediated Copper Uptake by* E. coli

A chalkophore-mediated copper uptake system is described in Chap. 3 for Gram-positive organisms, but the mechanism has recently been extended to *E. coli*. During urinary tract infection, *E. coli* expresses not only its prototype siderophore enterobactin, but a range of structurally different siderophores, presumably to acquire sparse iron [7]. It was argued that expressing different siderophores (stealth siderophores) would help to escape the scavenging of Fe^{3+}-siderophore complexes by the infection-associated protein siderocalin, which renders the iron inaccessible to *E. coli*.

The prominent virulence-associated siderophore yersiniabactin (Ybt) was first described in *Yersinia pestis*, but is often secreted alongside enterobactin by *E. coli* and *Klebsiella pneumoniae* during infection [8]. Ybt production is directed by the Yersinia high pathogenicity island, which encodes Ybt biosynthetic enzymes and membrane transporters. These genes are highly induced during urinary tract infections. It had been known for some time that some siderophores can also bind copper and Cu(II)-Ybt had been identified in urine from patients infected with Ybt-expressing *E. coli* [7].

Koh et al. [9] recently showed that Cu(II)-Ybt actually can serve as a nutritional source of copper for *E. coli* during urinary tract infections (Fig. 4.2). Cu(II)-Ybt is taken up into the periplasmic space by FyuA (ferric yersiniabactin uptake A), a TonB-dependent outer membrane transporter. TonB is an inner-membrane complex that couples the electrochemical potential across the cytoplasmic membrane to a variety of outer-membrane transporters [10]. It is a typical uptake mechanism for metal-siderophore complexes, but also for vitamin B_{12}, carbohydrates and other compounds. From the periplasmic space, Cu(II)-Ybt is transported across the cytoplasmic membrane by an ABC-type (ATP-binding cassette type) transporter, consisting of YbtP and YbtQ. The Cu^{2+} is converted to Cu^+, probably upon entering the reducing environment of the cytoplasm, which leads to Cu^+ dissociation from Ybt. This is supported by the inability to release redox-inactive Ga^{3+} from Ybt. Demetallated Ybt is then recycled by secretion through unknown transport systems. In a strain expressing a periplasmic copper amine oxidase, this activity was enhanced by the presence of the Ybt-system which could apparently contribute to the delivery of copper to this oxidase; this copper is thus routed through the cytoplasm.

The Ybt-directed copper uptake system of uropathogenic *E. coli* raises of course the question of how many of the over 500 known siderophores also, or even predominantly, act as chalkophores and how wide-spread such a mechanism is "in the wild". Most research on copper homeostasis was conducted on laboratory organisms under laboratory growth conditions, which are usually optimized for fast growth or

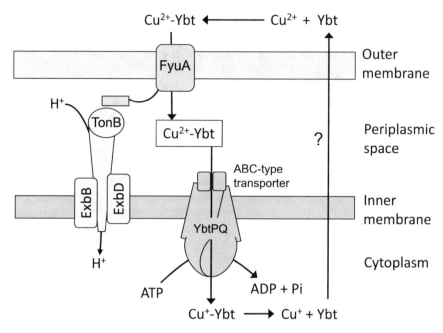

Fig. 4.2 Uptake of Cu(II)-Ybt directed by the Yersinia high pathogenicity island. Cu(II)-Ybt is transported across the outer membrane by FyuA (ferric yersiniabactin uptake A), driven by the TonB complex, involving TonB, ExbB, and ExbD. From the periplasmic space, Cu(II)-Ybt is transported across the cytoplasmic membrane by the ABC-type transporter, YbtPQ. In the cytoplasm, Cu(II) is reduced to Cu(I), which leads to its dissociation from Ybt and *apo*-Ybt is recycled

other experimental requirements. In addition, they are almost always monocultures, while in nature, bacteria always grow in complex communities. The Yersinia high pathogenicity island and similar structures are highly mobile and can jump between species. Conceivably, under real-life conditions, chalkophore-mediated copper uptake is the rule rather than the exception.

As a corollary to copper-sequestration by Ybt, this mechanism could also serve as a defense during infection. While hosts usually restrict metal ions during infection, the situation is different in the phagosomes of macrophages [11, 12]. Upon infection, the copper ATPase ATP7A is recruited to the phagosomal membrane and copper is actively pumped into the phagosome. The oxidizing environment and the activity of CueO-type multicopper oxidases would oxidize Cu^+ to Cu^{2+}, which binds to chalkophores or siderophores with high affinity.

4.1.3 Chalkophore Mediated Copper Uptake by Methanotrophs

Methanotrophic organisms require copper for the synthesis of membrane-bound methane monooxygenase (particulate, pMMO). They can take up copper with the help of methanobactins (Mbt). Mbt of *M. trichosporium* OB3b is a small, 1.2 kDa peptides of 8–10 amino acids that are heavily modified posttranslationally (Fig. 4.3a) [13]. Mbt is ribosomally synthesized as precursor proteins with a secretion signal that is cleaved. Next to the methanobactin structural gene, *mbtA*, there are up to a dozen or more additional genes, organized into one or two operons, including two putative biosynthetic genes, *mbnB* and *mbnC*, and additional biosynthetic proteins like amino-transferases, sulfotransferases and flavin adenine dinucleotide (FAD)-dependent oxidoreductases. In addition, there are genes for a conserved set of transporters, including multidrug exporters and TonB-dependent transporters, and a di-heme cytochrome *c* peroxidase. All these genes are probably involved in methanobactin synthesis, secretion, and uptake [14]. Methanobactins of varying chemical composition have been isolated from different methanotrophs [15].

Mbt is probably secreted into the periplasmic space by MbnM and across the periplasmic membrane by unknown proteins in the periplasmic membrane (Fig. 4.3b). Secreted Mbt binds Cu^{2+} from the environment with high affinity, followed by reduction of the bound copper to Cu^+. Cu^+-Mbt is actively taken up into the periplasmic space by a TonB-dependent transporter in the outer membrane (MbnT) by a process that is driven by the proton motive force [16]. Transport from the periplasm into the cytoplasm appears to be catalyzed by an ABC-type transporter and is powered by ATP. Copper-limitation stimulated Mbt secretion.

It has been suggested that methanobactins not only serve in copper acquisition, but could also provide protection against other heavy metals like mercury or gold, which are also complexed by these molecules but not taken up by cells. Furthermore, methanobactins could conceivably have a signaling function in bacterial communities [15]. Clearly, there is still much to be learned about methanobactins and chalkophores in general. However, in terms of copper acquisition, a new concept starts to emerge, namely, that many, if not most bacterial species can take up copper bound to small, organic ligands, or chalkophores, which have either been synthesized by the organism itself or by another species in the community. Like siderophores for iron acquisition, methanobactins can thus be shared in a bacterial community, that is, one species can utilize the methanobactin produced by another species [16]. *Methylococcus capsulatus* (Bath) was found to secrete a 60 kDa copper-binding protein, MopE, which is located on the cell surface, but also released into the medium in a truncated form. MopE binds one Cu^+ with high affinity and one Cu^{2+} with low affinity [17]. The role of MopE in *M. capsulatus* remains unclear, but the protein could conceivably function in the scavenging of copper outside the cell.

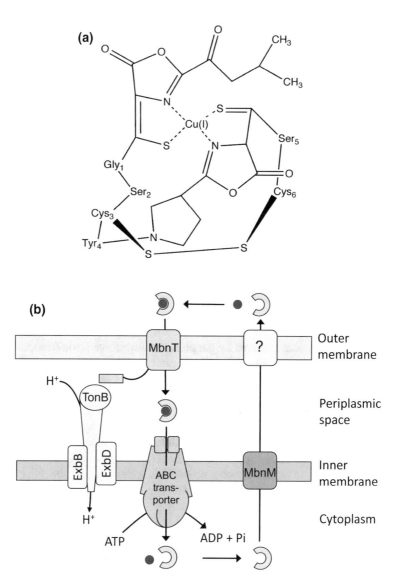

Fig. 4.3 Ligand-guided copper uptake by *M. trichosporium* OBb3. **a** Structure of the methanobactin of *M. trichosporium* OB3b, of molecular weight 1217 Da. **b** The chalkophore (brown semi-circles) may be secreted into the periplasmic space by MbnM and to the outside by an unknown protein in the periplasmic membrane. Cu^+ (blue spheres) is scavenged from the environment by the chalkophore and the Cu-chalkophore complex is actively taken up into the periplasmic space by a TonB-dependent transporter (MbnT). Uptake into the cytoplasm is probably accomplished by a chelate ABC-transporter. Copper could is released from Mbt either in the periplasmic space or in the cytoplasm

4.1.4 Other Proposed Copper Uptake Systems

There have been a number of reports of bacterial copper uptake systems. However, none of these systems has unequivocally been shown to catalyze copper uptake. Conceivably, these systems have different functions in copper homeostasis and further work will be required. See also Sect. 4.8. for additional information.

4.1.4.1 Copper Uptake by CopD-like Proteins

In *P. fluorescens* SBW25, CopD was proposed to mediate copper transport into the cytoplasm. The expression of the *copCD* operon is controlled by the CopRS two-component regulatory system [18]. Mutants devoid of CopCD or CopS were more tolerant to copper, while overexpression of CopCD reduced copper tolerance. From these observations, it was concluded that CopD is a copper uptake system. CopD is predicted to be a membrane protein with eight transmembranous helices, while CopC was shown to be a periplasmic protein with binding sites for Cu^+ and Cu^{2+} and presumably functions as a copper chaperone [19]. CopC from *Methylosinus trichosporium* OB3b, however, only features a single Cu^{2+} binding site [20]. *Bacillus subtilis* YcnJ is a CopCD fusion and has also been implicated in copper transport into the cytoplasm, chiefly based on the study of gene regulation [21]. Overexpression of the *copCD* genes in *Pseudomonas syringae* also led to copper accumulation and hypersensitivity to copper [22]. While the evidence for a role of CopD in copper uptake appears convincing, alternative roles for this protein should be considered. Copper uptake by bacteria is usually assessed by measuring total copper in cells; copper accumulation only in the periplasmic space would not easily be detected and could skew the experimental results. As an alternative function for CopD, transport of copper to the periplasmic space for the metallation of cuproenzymes should also be considered. Indeed, in *Methylosinus trichosporium* OB3b, CopCD did not appear to be critical for copper uptake or the switch between the expression of copper-containing or iron-containing methane monooxygenase [23].

4.1.4.2 CtaA and PacS of *Synechocystis* PCC680

Tottey et al. [1] claimed that in the photosynthetic organism *Synechocystis* PCC680, copper is supplied to the lumen of the thylakoids by two P1B-type coper ATPases, CtaA and PacS. It was hypothesized by that CtaA serves in the import of copper into the cytoplasm and PacS then transports the copper from the cytoplasm into the thylakoid lumen for use in plastocyanin, a copper containing electron transfer protein. The conclusion was based on very indirect evidence, mainly based on an early study in which PacS was localized to the thylakoid membrane in the cyanobacterium, *Synechococcus* PCC7942 [24]. However, this was based on a crude membrane separation without the use of maker enzymes (thylakoid membranes were identified as being green). We now know that the biogenesis of thylakoid membranes in cyanobacteria is complex and that they form a continuum with the cytoplasmic membrane, so there is no need to transport copper across the cytoplasm [25, 26] (see Sect. 4.8).

4.1.4.3   CcoA of *Rhodobacter capsulatus* MT1131

It was proposed that CcoA, a major facilitator superfamily (MFS)-type transporter, imports copper into the cytoplasm for the biogenesis of the cytochrome cbb_3-oxidase in *Rhodobacter capsulatus* MT1131 [27, 28]. This was based on the following, indirect evidence: a mutant in CcoA was defective in cytochrome cbb_3-oxidase and exhibited less copper accumulation, *measured on whole cells* (not just the cytoplasm!). Additional support for the hypothesis was drawn from the observation that spontaneous revertants of the $\Delta ccoA$ mutant, which again expressed functional cytochrome cbb_3-oxidase were defective in CopA, the major copper exporter. These revertants were copper sensitive, had raised copper levels, and could again produce functional cytochrome cbb_3-oxidase [28]. Single $\Delta copA$ mutants also had elevated copper levels, were copper sensitive, and produced functional cytochrome cbb_3-oxidase. So elevated copper levels restored cytochrome cbb_3-oxidase expression in the absence of CcoA by an unknown rescue pathway. An alternative explanation of CcoA function should not be disregarded: that it metallates cytochrome oxidase by transporting cytoplasmic copper to the periplasm, and not vice versa. The experiments could not differentiate between cytoplasmic and periplasmic copper levels–only whole-cell copper levels were measured.

Additional evidence taken in support of CcoA being a copper importer was derived from its expression in yeast. In contrast to bacteria, eukaryotic cells depend on copper import across the cytoplasmic membrane, accomplished by Ctr-type transporters (see Ref. [29] for review). *Schizosaccharomyces pombe* defective in the plasma membrane copper importers Ctr4 and Ctr5 is defective in copper uptake and lacks copper-containing cytochrome c oxidase. It can thus not grow on non-fermentable carbon sources like glycerol or ethanol. The *S. pombe* copper transporter Mfc1 is expressed only during spore formation and is localized in the forespore membrane of ascospores, where it transports copper into the forespore [30, 31]. If Mfc1 is continuously expressed from an engineered promoter, it restores cytochrome c oxidase activity and thus growth on glycerol (if the growth medium is supplemented with $2\,\mu M$ copper) in the absence of functional Ctr4 and Ctr5. It was shown that under these conditions, the copper primarily resides in the forespore and could apparently reach other sites, such as for the synthesis of cytochrome c oxidase, via alternative pathways. Like Mfc1, *R. capsulatus* CcoA could also restore growth of the *S. pombe* $\Delta ctr4/\Delta ctr5$ mutant on glycerol [32]. CcoA exhibits 37% sequence similarity to Mfc1 and it appears likely that CcoA is also localized in the forespore and transports copper into this compartment, from where copper could reach cytochrome c oxidase by an alternative route. By site-directed mutagenesis, amino acid residues of CcoA that are required for copper accumulation by whole cells were identified [33]. So it is undisputed that CcoA is a copper transporter but in *S. pombe*, it restored cytochrome oxidase metallation by delivering copper to the forespore, and not directly to the cytoplasm.

Therefore, while many experimental observations could be interpreted as CcoA delivering copper to the cytoplasm, there could be an alternative explanation: CcoA normally transports cytoplasmic copper to the periplasmic space to provide it for the assembly of cytochrome cbb_3-oxidase. This results in increased periplasmic copper, possibly compartmentalized, and thus to an *apparent* increase in whole-cell copper. This hypothesis is supported by the observation that an *S. pompe* $\Delta ccoA/\Delta copA$ double mutant which lacks the major copper export ATPase, CopA, can still make functional cytochrome oxidase, presumeanly because higher cytoplasmic copper allows copper to reach the periplasm by an alternative route. It will be important to clearly differentiate between cytoplasmic and periplasmic copper levels in these experiments, and future work will show which of the two hypothesis are correct.

4.2 Copper Chaperones and Synthesis of CopZ in *E. coli* by Ribosome Slippage

CopZ-type copper chaperones have been discussed in detail in Chap. 3. There has been a long and unsuccessful search for a CopZ-type copper chaperone gene in *E. coli*. These chaperones consist of only around 70 amino acids and the corresponding genes are thus only a bit over 200 nucleotides in length. Such small genes are often missed by automatic annotation programs, particularly if the gene does not start with the methionine initiator codon ATG. However, the puzzle of the missing *E. coli* CopZ gene was recently solved by two labs [34, 35]. It was found that by a mechanism of programmed ribosomal frameshifting, transcription of the CopA copper ATPase gene is halted after translation of the first MBD (Fig. 4.4). This produces a peptide of 69 amino acid which contains the MBD1 domain and can thus work like a CopZ-type chaperone.

In addition to the programmed ribosomal frameshifting, the production of a short mRNA transcript of the *copA* gene was observed [35]. Most of the copper chaperone apparently derived from this highly abundant transcript, rather than from the full-length *copA* transcript, but the mechanism whereby this short transcript is produced remains unclear. Overall, the amount of transcribed CopZ exceeded that of CopA and appears sufficient to fulfil the chaperone function.

Eliminating chaperone expression in *E. coli* by introducing a corresponding muta-tion in the N-terminus of the *copA* gene did not result in a phenotype. Only in a mutant devoid of GSH synthesis, an increased resistance to a brief copper shock could be observed [35]. This fits a general pattern that has been observed, namely that CopZ-like chaperones appear dispensable, at least under laboratory conditions. But the extreme conservation of CopZ chaperones among bacterial species, and in fact all forms of life, demonstrate that this function is far from dispensable. As pointed out already in Chap. 3, laboratory conditions fall short of reflecting the 'real-life' sit-

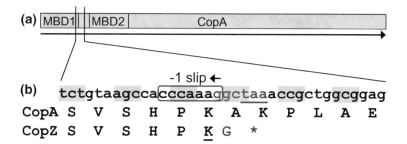

Fig. 4.4 Generation of CopZ in *E. coli* by programmed ribosomal frameshifting. **a** *CopA* gene with the two MBDs. **b** Enlarged region between MBD1 and MBD2, showing the nucleotide sequence with the normal reading frame punctuated by green boxes and the 'slippery' sequence 5′-cccaaag boxed in blue (followed by a pseudoknot [34]). Slippage of ribosomes by −1 after the underlined K-residue results in translation of the nucleotides depicted in red, ggctaa, encoding Glycine and a stop

uation and it will be important to investigate bacterial copper homeostasis also in natural habitats and in bacterial communities.

Structures which would lead to programmed ribosomal frameshifting were not only detected in the N-termini of many bacterial copper ATPases, but also in the N-terminus of the human Wilson copper ATPase, ATP7B [34]. This enzyme catalyzes copper secretion from the liver into the bile and features six N-terminal MBDs [36]. Premature termination of translation is predicted to occur between MBD2 and MBD3. This would result in the production of a truncated protein composed of MBD1 and MBD2. That such a mechanism which produces additional copper chaperones is conserved among species shows that it must have an evolutionary advantage.

4.3 Copper ATPases

A MerR-type copper-responsive transcriptional activator, CueR, regulates the expression of two genes important for copper homeostasis: the CopA copper efflux ATPase and the periplasmic CueO multi-copper oxidase (see Sect. 4.4.) [37]. *E. coli* appears to belong to a minority of bacterial species, including Gram-positives, that only possesses a single copper ATPase. Most bacteria contain two (and some even more) P1B-type Cu-ATPases with different efflux rates and these enzymes have different physiological roles: copper ATPases with high efflux rate serve in the detoxification of the cytoplasm, while alternative ATPase with low efflux rates are engaged in the metallation of cuproenzymes in the periplasmic space (cf. Sect. 4.8.).

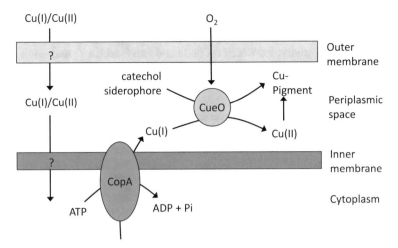

Fig. 4.5 Periplasmic copper toxicity and detoxification by CueO. Toxic Cu(I) entering the periplasmic space either from the outside or from the cytoplasm by extrusion of Cu(I) by CopA can be oxidized to less toxic Cu(II) by CueO. Alternatively, CueO can oxidize catechol siderophores and the resultant pigment can oxidize Cu(I) to Cu(II) and bind Cu(II). Molecular oxygen provides the oxidizing equivalents for CueO

4.4 Extracellular/Periplasmic Copper Oxidation or Reduction

4.4.1 CueO-Catalyzed Copper Oxidation

CueO of *E. coli* is a multicopper oxidase (MCO) that has a robust cuprous oxidase activity which can contribute to copper resistance [38, 39]. *E. coli* defective in CueO becomes more sensitive to copper toxicity and this effect is more pronounced in synthetic growth media compared to complex growth media like Luria-Bertani broth [40]. One possible contribution of CueO to copper tolerance lies in the oxidation of toxic Cu^+ to Cu^{2+} (Fig. 4.5) [41, 42]. Cu^+ can lead to the generation of very damaging reactive hydroxyl radicals. Tiron, a superoxide quencher, was able to suppress the phenotype of a $\Delta cueO$ mutant, suggesting that the generation of oxidative stress in the periplasm is a toxicity factor (see also Chap. 2) [40]. It should be emphasized here that superoxide production is not the a priori copper toxicity mechanism in wild-type *E. coli*, but it can be observed in $\Delta cueO$ mutants. As such, it could thus be considered an 'accessory' toxicity mechanisms, meaning that such a mechanisms is not apparent in wild-type cells, but can become significant in a mutant cell. A number of other such mechanisms have been described, but they will not be discussed in this book.

A second mechanism by which CueO can contribute to copper resistance is by the oxidation of siderophores and other phenolic compounds to their polyphenols.

Catechols like the siderophore enterobactin can reduce Cu^{2+} to toxic Cu^+, thereby fueling the production of reactive oxygen species. The oxidation of catechols results in the formation of pigments, which on one hand can bind Cu^{2+}, and on the other hand can also oxidize Cu^+ to Cu^{2+} [39]. So CueO is at the center of a network of beneficial and detrimental reactions. CueO requires molecular oxygen and is thus a defense mechanism primarily active under aerobic conditions. Under anaerobic conditions, the CusCFBA copper extrusion system becomes vital to prevent toxicity of periplasmic copper. A similar role in copper tolerance was also demonstrated for the MCOs of *Brucella melitensis* (BmcO), *Desulfosporosinus* OT, and *Mycobacterium tuberculosis* (MmcO) [43–45]. In pathogenic bacteria, MCOs may be an important virulence factor by supporting survival in phagosomes [12].

In activated macrophages, the eukaryotic copper ATPase, ATP7A, traffics from the *trans*-Golgi network to phagosomes and raises the phagosomal copper concentration [46, 47]. Such a copper-up response may not be limited to macrophage, but could occur in a range of cell types in response to internalized bacteria or other factors such as inflammation [29, 48]. MmcO has been shown to be required for copper resistance by *M. tuberculosis*, presumably by oxidizing Cu^+ to less toxic Cu^{2+}, thus aiding survival in the human host [45]. On the other hand, survival of *Salmonella* Typhimurium deficient in CueO was not attenuated in macrophages, but was reduced in liver and spleen [49]. The MCO of *Brucella melitensis*, BmcO, did also not affect survival in macrophages [43].

Clearly, CueO-type MCO contributes to bacterial copper resistance in culture, but it may not be a key factor in all species for survival in phagosomes [50]. This may have multiple underlying reasons. For one, the phagosomal interior may be sufficiently oxidizing for spontaneous copper oxidation to occur. Secondly, the primary role of CueO-like enzymes in vivo may not always be the oxidation of copper, but the oxidation of iron or organic substrates [12]. In a number of species, e.g. *Pseudomonas aeruginosa*, *Legionella pneumophila*, or *Desulfosporosinus* OZ, CueO-like enzymes were shown to also possess iron oxidase activity, suggesting that they may have a role iron acquisition [44, 51, 52]. So the connection of MCOs to copper tolerance and virulence is not straight-forward and several mechanisms may contribute to phenotypic observation, namely cuprous oxidase activity, ferrous oxidase activity, phenolic oxidase activity, copper complexation by polyphenols, and iron acquisition.

4.4.2 Copper Reduction Systems

In *E. coli*, a copper reductase system has been described. It has originally been ascribed to the NADH dehydrogenase NDH-2 [53], but later work showed that copper reduction is mainly due to ubiquinone in the cytoplasmic membrane, and only to a minor extent to NDH-2 in a quinone-independent way [54]. The major copper reductase activity resembles the non-enzymatic copper reduction observed in *L. lactis* at the cytoplasmic membrane [55] (see Sect. 3.1). Since these reactions do not involve enzyme catalysis, they are obviously fortuitous. As such, they are likely to occur on the cytoplasmic membranes of all bacteria and may well lead to artifacts in certain type of experiments.

4.5 The ComC/ComR Outer Membrane System

Using a biosensor for cytoplasmic copper, Mermod et al. [56] screened for *E. coli* mutants which would have lower influx of copper into the cytoplasm. This led to the identification of the outer membrane protein ComC. Transcription of *comC* is repressed by CopR under low copper conditions. Growth of *E. coli* in high copper induces ComC, which is made as a 12.4 kDa precursor. Upon export to the outer membrane, it is processed to the mature protein of 11 kDa. Increased ComC expression results in a lower permeability of the outer membrane to copper. Predicted ComC-like proteins could be identified in all Gram-negative organisms analyzed, but were absent in Gram-positive bacteria, in line with their lack of an outer membrane.

ComC could reduce the copper permeability of the outer membrane in two ways, either by directly making the lipid bilayer less copper-permeable, or by interacting with proteins such a porins, which allow copper to access the periplasmic space. In either case, these findings shown that copper does not have unrestricted access to the periplasmic space, such as by uncontrolled diffusion through porins.

The second point of interest raised by this study is the ComR repressor. It is the first TetR-like repressor which has been linked to copper as the inducer. TetR-like repressors form a large, diverse group and over 100 crystallographic structures are known [57]. However, only six of them so far have a proven biological inducer. The *E. coli comR* gene is distant to *comC* on the chromosome and is not regulated by copper. ComR could well regulate additional genes in *E. coli* and other Gram-negative organisms in a copper dependent manner.

The mechanism of regulation of ComC by ComR is not understood in molecular detail. Like most repressors, ComR is a homodimer consisting of two 26 kDa monomers and features a helix-turn-helix DNA binding domain and a putative copper binding site formed by C117, M119, and Q179 [56]. In the absence of copper, ComR binds to the *comC* promoter and suppresses transcription. Copper releases ComR from the promoter, allowing transcription to proceed. In addition, expression of ComR appears to be co-regulated by RpoE, a general regulator of extracytoplasmic function proteins, and the cAMP response protein CRP. However, induction of *comC* by copper via ComR is much more pronounced than the effects of RpoE and CRP. Other TetR-like copper-responsive repressors have so far not been describe, but will certainly exist.

4.6   Copper Transport Across the Outer Membrane

4.6.1   The E. coli *Cus-System and Its Regulation*

While Gram-positive bacteria with their single cell membrane can rely on a single copper efflux ATPase for copper tolerance, Gram-negative organisms harbor a second copper export system, CusCFBA, for copper transport across the outer membrane. These efflux systems are members of the tripartite resistance–nodulation–cell division (RND) family of transporters, consisting of CusCBA in *E. coli* [58]. CusF is a periplasmic copper chaperone that scavenges copper in the periplasm and conveys it to the CusCBA transporter (cf. Fig. 4.1).

The CusCFBA copper transport system has been well characterized in *E. coli* (see Ref. [59] for review). It is a big complex of stoichiometry $CusC_3B_6A_3$ that is anchored in the cytoplasmic membrane by CusA, spans the periplasmic space, and connects to the outer membrane via CusC. The crystal structure of a $CusB_6$–$CusA_3$ complex shows that 12 transmembranous helixes of each CopA monomer form the anchor in the cytoplasmic membrane [60]. Six CusB subunits sit on top of the CusA trimer and a CusC trimer forms an ion channel that is anchored in the outer membrane by a β-barrel structure [61]. Substrate binding occurs via CusA, where Cu^+ (or Ag^+) could enter via a periplasmic cleft or from the cytoplasm. However, it is believed that CusCBA predominately or even exclusively transports periplasmic copper across the outer membrane [62]. Transport is driven by the proton-motive force across the inner membrane.

Copper in the periplasmic space can originate either from the outside by leakage across the outer membrane or from the cytoplasm, by transport into the periplasmic space by CopA. It has been shown that CopA can directly donate Cu^+ to *apo*-CusF; the resultant Cu^+–CusF complex in turn can donate Cu^+ to CusB of the CusCBA complex for transport across the outer membrane [58, 63]. This allows precise targeting of copper to the cellular export pathway. Copper handling in the periplasm may in fact attain a high degree of complexity. In *Sinorhizobium meliloti*, five different Cu^+-ATPases may pump copper into the periplasmic space, where two putative CusF chaperones and two SenC-like proteins participate in the routing of Cu^+ for export, metallation of cytochrome *c* oxidase, and possibly other functions [64].

Expression of the *cusCFBA* operon is under the control of the CusRS two-component regulatory system, a type of regulatory system exclusively found in Gram-negative organisms [65]. CusS is a sensor kinase that traverses the cytoplasmic membrane and senses copper in the periplasmic space. When copper binds to CusS, a conserved histidine of the protein autophosphorylates, followed by phosphotransfer to a conserved aspartic acid residue on CusR, the cytoplasmic response regulator. Phosphorylated CusR ultimately activates transcription of the *cusCFBA* and the *cusRS* operons. Similar two-component regulators have been shown to also control the periplasmic copper export machinery in widely different species like *Synechocystis* sp. PCC 6803, *Corynebacterium glutamicum*, or *Pseudomonas fluorescens* and may thus be universal [66–68].

The CueR and the CusRS regulons differ in their response to copper. CueR, which controls CueO and CopA expression, is induced by relatively low external copper concentrations while CusRS, which controls CusCFBA expression, requires higher external copper concentrations and may be most important under anaerobic conditions. So the fate of copper which has been exported to the periplasm by CopA or entered from the outside of the cell maybe twofold, either (i) oxidation by CueO to less toxic Cu^{2+}, primarily under aerobic conditions, or (ii) transport of the Cu^+ to the extracellular space by CusCBA, primarily under anaerobic conditions.

4.6.2 The Mycobacterial MctB System

Mycobacteria will be discusses briefly here because they are unique among bacteria. They feature a complex cell wall containing unusual lipids and thereby forming a kind of outer membrane [69]. So the classification as either Gram-negative of Gram-positive organisms is controversial, but phylogenetically, they are more closely related to Gram-negative organisms. Like in Gram-negative organisms, the outer membrane struture functions as a barrier to toxic substances. It also contains porins that are required for the uptake of essential nutrients and may also allow copper to get access to the cytoplasmic membrane [70]. In *Mycobacterium tuberculosis*, excess cytoplasmic copper induces the *cso* operon, which is under the control of the copper-inducible CsoR repressor [71]. The operon encodes CtpV, which is P1B-type copper ATPase which pumps cytoplasmic copper across the cytoplasmic membrane. Excess copper also induces two proteins of unknown function, Rv0970 and Rv0968. *M. tuberculosis* and other mycobacteria do not possess a CusCFBA system. Instead, MctB is required to transport the periplasmic copper to the cell exterior. Mutation of the *mctB* gene dramatically reduces the virulence of *M. tuberculosis*, highlighting the important role of copper in killing bacteria in phagosome [12, 72]. A second copper-inducible repressor, RicR, controls its own expression and the expression of the five monocistronic genes *mymT*, encoding a metallothionein which can sequester cytoplasmic copper, *mmcO*, encoding a CueO-type MCO, and three genes of unknown function, *lpqS*, *rv2963*, and *socAB* [73]. Clearly, further work will be required to reach an understanding of copper homeostasis in mycobacteria at the level of that we have for *E. coli* or *E. hirae*.

4.7 Copper Buffering and Storage

4.7.1 Copper Buffering by Glutathione (GSH)

GSH is present in the cytoplasm of most Gram-positive bacteria [74–77]. GSH is in equilibrium with the oxidized, dimeric form, GSSG. The GSH/GSSG redox couple helps to maintain a negative cytoplasmic redox potential in growing cells [78] and also keeps cytoplasmic copper in the reduced Cu^+ form. It was shown that under anaerobic conditions, GSH enhanced copper toxicity through the formation of GS-Cu-SG complexes (two GSH molecules complexing one Cu), which then could donate copper to enzyme sites reserved for other transition metals [79] (see Sect. 3.6). This was so far only shown in Gram-positive *Lactococcus lactis* and it is not clear if this is a general mechanism, i.e. if GSH would also enhance copper toxicity in anaerobically grown *E. coli*.

Under aerobic conditions, *E. coli* devoid of the GSH biosynthetic pathway is considerably more sensitive to killing by copper than wild-type cells [80]. This is particularly apparent in the absence of a functioning copper export ATPase. *Streptococcus pneumoniae* cannot synthesize GSH, but can take it up via an ABC-type transporter [81]. When GSH was supplied to growing *S. pneumonia*, the cells became considerably more resistant to copper, but also to Cd^{2+} and Zn^{2+}. In *E. coli*, GSH was also shown to increase resistance towards Hg^{2+} and arsenite (AsO^{2-}) [82]. Clearly, GSH can protect cells from heavy metal toxicity under aerobic conditions, especially if they are compromised in other ways, such as by mutation or other stresses conditions. But it must be emphasized that wild-type cells are quite well equipped to deal with various 'every-day' stress conditions and that GSH and similar systems should be viewed as back-up systems under stress conditions.

4.7.2   Copper Buffering by Metallothioneins

Metallothioneins are best known from eukaryotic cells. They are cysteine-rich proteins of 4–14 kDa designed mainly for the binding of Zn^{2+}, Cd^{2+}, and Cu^+ [83]. Metallothionein-like proteins are rare in prokaryotes and the only well-characterized metallothionein-like protein is SmtA from Gram-negative *Synechococcus* PCC7942 [84]. SmtA is induced by heavy metal ions via the SmtB repressor. Similar genes can be found in other organisms with bioinformatics tools, but no SmtB-homologue has been characterized so far.

In *Mycobacterium tuberculosis*, a 4.9 kDa cysteine-rich protein which functions as a metallothionein was identified and termed MymT (mycobacterial metallothionein) [85]. The protein is induced by copper or cadmium and helps to protect the organism against copper toxicity. In vitro, MymT binds up to six copper ions in a solvent shielded environment, which is typical for metallothioneins. The amino acid sequence of MymT was found to be highly conserved among pathogenic mycobacteria.

The methanotroph *M. trichosporium* OB3b possesses three copper storage proteins: Csp1, Csp2, and Csp3. Csp1 and 2 have predicted twin arginine translocation signals and thus appear to be exported from the cytoplasm, while Csp3 is cytosolic [86]. Csp3 was also discovered in Gram-positive *B. subtilis* (see Chap. 3). The crystal structures of *B. subtilis* and *M. trichosporium* Csp3 have been solved and revealed a tetramer consisting of 4-helix bundle monomers. The tetramer can bind up to 80 copper atoms via thiolate groups [86–88].

The ability of bacteria to store significant amounts of copper intracellularly is a novel aspect. When and how this copper will be used for the biosynthesis of cuproenzymes remains to be shown.

4.7.3 Copper Tolerance Mediated by Polyphosphate

Polyphosphate is a linear polymer of orthophosphate residues linked by high-energy phosphoanhydride bonds. These polymers typically consist of hundreds of phosphate residues and have been found in all cell types. Polyphosphates are involved in cell growth, response to stresses, and the virulence of pathogens (see Ref. [89] for review). The synthesis of polyphosphate is catalyzed by polyphosphate kinases 1 and 2 (PKK1, PKK2), which are highly conserved in many bacterial species, but absent in yeast and animal cells. Mutants lacking PPK1 and/or PPK2 are defective in various functions, such as quorum sensing, motility, biofilm formation, or virulence. Exopolyphosphatases (PPX) processively hydrolyze polyphosphate under release of inorganic phosphate (Pi). *E. coli*, other Eubacteria and Archaea contain two PPX, PPX1 and GPPA (PPX2). Intimately connected to polyphosphate synthesis and hydrolysis are of course the various phosphate transporters that occur in bacteria.

Limited work has been carried out on the connection between polyphosphate and heavy metal tolerance in bacteria. A copper-resistant strain of *Anabaena variabilis* was shown to contain more polyphosphate bodies and had higher internal phosphate levels than the sensitive strain [90]. Also, it was shown that polyphosphate metabolism in *E. coli* affected cadmium tolerance (copper tolerance was not tested but was most likely similarly affected) and PKK and PPX mutant strains were less cadmium tolerant [91]. It was proposed that in the presence of heavy metals, polyphosphate is degraded to orthophosphate by PPX and metal–phosphate complexes are transported out of the cell by an inorganic phosphate transporter [92].

In *Acinetobacter johnsonii*, it was shown that metal-phosphate efflux via the Pit system produces a proton motive force that can drive energy-requiring processes [93]. Thus, copper tolerance can be a concerted effort by copper secretion by ATPases and Cu-phosphate efflux to detoxify the cytoplasm [94].

In copper resistant strains of *Acidithiobacillus ferrooxidans*, it was similarly shown that copper, cadmium and zinc stimulated PPX activity and it was similarly proposed that this results in metal detoxification by the transport of metal-phosphate complexes out of the cell [95]. *Sulfolobus metallicus* accumulates polyphosphate granules. When exposed to high extracellular copper, PPX was induced and the polyphosphate levels decrease with increasing copper concentrations [96]. At the same time, phosphate efflux was stimulated. These results again suggest that polyphosphates mediate a metal tolerance mechanism.

Taken together, these and other finding support a model whereby copper tolerance can be accomplished or supported by the degradation of polyphosphate and

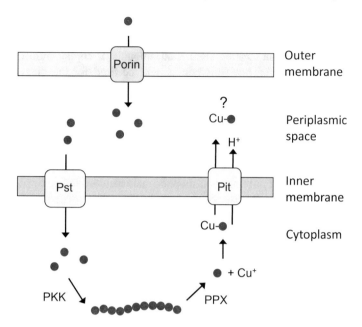

Fig. 4.6 Polyphosphate in copper resistance. In *E. coli*, copper entering the periplasmic space *via* porins is taken up into the cytoplasm by Pst or another phosphate transporter and is polymerized to polyphosphate by PKK1 or PKK2 (PPK). PPX1 or PPX2 (PPX) can release phosphate from the polyphosphate chains upon demand. If excess copper is present, copper-phosphate complexes form and leave the cytoplasm *via* PitA or PitB (Pit), which are low-affinity phosphate-H$^+$ symporters that can operate in either direction. What happens to copper-phosphate in the periplasmic space is not known. Copper could conceivably be transported across the outer membrane by the Cus-system. The red dots symbolize phosphate ions

the subsequent export of copper-phosphate complexes out of the cell by phosphate transporters (Fig. 4.6). This can be considered as an accessory system for copper tolerance, in addition to copper export by the copper ATPase. In the periplasm, Cu$^+$ exported as a Cu$^+$-phosphate complex could be detoxified by oxidation of the Cu$^+$ to less toxic Cu^{2+} by CueO or transport across the outer membrane by the Cus-system. Copper tolerance by copper-phosphate export has unfortunately not been taken into due considerations in most studies on copper homeostasis.

4.8 Copper-Loading of Cuproenzymes

A general concept appears to emerge for the metallation of cuproenzymes. It is based on the following steps: (i) copper enters the cytoplasm, either adventitiously as copper bound to organic substrates, or by active chalkophore-mediated uptake, (ii) excess cytoplasmic copper is expelled by specialized 'bilge pump' copper ATPases, and

(iii) copper for enzyme metallation is transported across the cytoplasmic membrane by 'metallation pumps' or transporters, which deliver the copper directly or via a chaperone to the enzyme to be metallated. Passage of the copper through the cytoplasm provides a check-point that only Cu^+ ions, and not other metal ions, are delivered to a cuproenzyme. Of course nature is very diverse and there may be variations to the theme. But given the high conservation of the copper homeostatic machinery, the concept may hold true for most organism. On the other hand, there are a number of reports of copper import proteins involved in the metallation of cuproenzymes, but we here favor the view that copper is never transported into the bacterial cytoplasm. Time will show which concepts is correct.

In support of this concept, most microorganisms possess two or more copper export ATPases. Metallation pumps are able to very specifically deliver Cu^+ to cuproenzymes and thus prevent inadvertent incorporation of another metal ion. Reports on ATPase-catalyzed copper *uptake* were based on indirect observations, never direct copper transport, and were single reports without any follow-up work. Other proteins which were postulated to serve in copper uptake may in fact serve in copper export for the metallation of cuproenzymes (see Sect. 4.1.) It is conceivable that some of the earlier interpretations of experimental observations are mis-interpretations. In particular, the concept that copper delivery to thylakoid- or chromatophore-localized plastocyanin and cytochrome oxidase requires copper transport into the cytoplasm and from there into the chromatophores might be wrong (e.g. Ref. [1]). In was recently shown that the chromatophores of *R. sphaeroides* form a continuum with the cytoplasmic membrane and that the lumen of the chromatophores is connected to the periplasmic space, which is most likely the site for the metallation of all cuproenzymes [25, 26, 97].

4.8.1 Copper Loading of the CueO Multicopper Oxidase

The MCO CueO is secreted into the periplasm to oxidize the more membrane-permeable and toxic Cu^+ to Cu^{2+}. CueO itself requires copper for activity. It is transported by the twin-arginine translocation (Tat) pathway across the cytoplasmic membrane to the periplasmic space. The Tat system is generally assumed to transports folded proteins. However, CueO of *E.coli* was shown to be secreted into the periplasm as an *apo*-protein [98]. CueO is thus the first example for a Tat substrate that is secreted in the apo-form and cofactor assembly apparently takes place in the periplasmic space. Periplasmic *apo*-CueO could also be activated in vitro by the addition of copper.

4.8.2 Copper Loading of Cu,Zn-Superoxide Dismutase (SOD)

In bacteria, Cu,Zn-superoxide dismutases (SOD) are periplasmic or extracellular enzymes which protect cells against oxidative stress. In *Salmonella enterica* serovar Typhimurium, it has been shown that either one of two copper ATPases, CopA or GolT, is needed to activate periplasmic Cu,Zn-superoxide dismutase SodCII [99]. $\Delta copA/\Delta golT$ double-mutant cannot synthesize active SodCII. However, the enzyme contains zinc and can be activated in vitro by supplementing copper. Single $\Delta copA$ or $\Delta golT$ mutants, on the other hand, produce active Cu,Zn-SodCII. This shows that either one of the two P1B-type copper ATPases can supply copper to the periplasmic space for enzyme metallation. GolT has originally been identified as an ATPase involved in gold resistance, but can apparently also (or primarily) transport copper [100, 101].

For copper delivery to *Salmonella* Typhimurium SodCII, a small periplasmic protein, CueP, is also required [99]. This protein is co-regulated with CopA by the copper-inducible repressor CueR and is also involved in copper tolerance by cells [102]. CueP is a homodimer of two subunits of 158 amino acids. The monomer exhibits an N-terminal domain with an $\alpha\beta\beta\beta\alpha$-fold and the larger C-terminal domain is folded in a six-stranded β-sheet [103]. CueP apparently functions as a periplasmic copper chaperone to metallate SodCII. It is interesting to note that *E. coli* possesses a SodCII orthologue, but no protein resembling CueP. Conceivably, the *E. coli* enzyme acquires its copper via the CusF periplasmic chaperone. Clearly, it will be most interesting to learn more about the function of *Salmonella* CueP and the metallation of SodCII in *E. coli*.

4.8.3 Copper Loading of Cytochrome Oxidase

Cytochrome c oxidases (COX) are ubiquitous multi-subunit enzymes at the end of the respiratory chains of prokaryotic and eukaryotic cells. There are three types of COX enzymes: aa_3-type, ba_3-type, and cbb_3-type, the latter only occurring in bacteria [104]. Copper delivery and insertion into these different types of COX requires different assembly factors. All three type of COX possess an intramembranous catalytic subunit I, which contains a low-spin heme and a high-spin heme-Cu_B binuclear center. The aa_3-type COX and some ba_3-COX contain a second, binuclear Cu_A center in a surface-exposed, hydrophilic domain of subunit II. The delivery and insertion of Cu into the Cu_A and Cu_B centers follow distinct pathways.

The existence of copper ATPases specifically involved in the metallation of cuproenzymes is an emerging principle. Most bacteria produce at least two copper ATPases [105]. In *Pseudomonas aeruginosa* it was demonstrated that the copper ATPase CopA1 confers copper tolerance to the cells, while CopA2 is required for cytochrome oxidase assembly [106–108]. CopA2 did not convey copper tolerance to

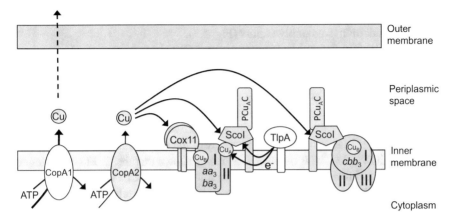

Fig. 4.7 Schematic drawing of the assembly of the copper centers of cytochrome oxidases. CopA1 is a high-rate copper ATPase responsible for copper tolerance, while CopA2 (CcoI, CtpA) is required for delivery of cytoplasmic copper to cuproenzymes. Cox11 (CoxG, CtaG) participates in the formation of the Cu_B center in aa_3-COX and ba_3-COX. ScoI (Sco1, SenC, PrrC) participates in the assembly of both, Cu_A and Cu_B sites and interacts with PCu_AC (PcuC, PccA). TlpA directly interacts with both, ScoI and subunit II of COX to reduce critical cysteine residues to allow the binding of copper. The three different types of COX are shown as aa_3 and ba_3 (depicted as one enzyme) and cbb_3, and their respective Cu_A and Cu_B centers are indicated. The Roman numerals designate the COX subunits

the organism and, conversely, CopA1 could not serve in cytochrome oxidase assembly. Also, a copper-sensitive $\Delta copA$ E. coli mutant was complemented by CopA1, but not by CopA2. Both, CopA1 and CopA2 apparently export copper to the periplasm, but CopA1 exhibits a ten-fold higher rate of copper transport than CopA2 (Fig. 4.7).

In *Rhodobacter capsulatus*, the assembly of the cytochrome cbb_3 oxidase requires the *ccoGHIS* gene cluster [5]. Based on sequence similarity to other proteins, *ccoI* encodes a P1B-type ATPase, most likely a copper ATPase. It was found that exogenously added copper cannot complement the defect in cytochrome oxidase assembly of a *ccoI* knock-out mutant. This suggests that direct copper delivery to cytochrome oxidase by CcoI is required for cytochrome cbb_3 oxidase biogenesis. Similarly, in *Rubrivivax gelatinosus*, the CcoI analogue, CtpA, was shown to be involved in cytochrome cbb_3 oxidase assembly [6]. CtpA is a P1B-type ATPases closely related to bacterial CopA-type copper exporting ATPases. Inactivation of CtpA resulted not only in cytochrome oxidase deficiency, but also in decreased activity of the cuproenzyme NosZ, which is located in the cytoplasmic membrane and functions as a nitrous oxide reductase [6]. Thus, CtpA is a copper ATPase that supplies copper to the periplasmic space specifically for the metallation of copper-requiring enzymes. Disruption of the *ctpA* gene did not result in enhanced sensitivity of *R. gelatinosus* to excess copper, so CtpA is not required for copper resistance, but is dedicated to the assembly of membrane-bound and periplasmic cuproenzymes, while a CopA-type copper ATPase is required to convey copper tolerance. Copper pumped to the periplasmic space by CopA2-type ATPases is delivered to the different copper sites of COX by additional assembly factors (Fig. 4.7).

Cox11 (CoxG in *Bradyrhizobium japonicum*, CtaG in *P. denitrificans*) is a universal COX assembly factor for Cu_B sites of aa_3- and ba_3-type COX, but not of cbb_3-type COX, in bacteria as well as mitochondria [109, 110]. Cox11 is a 28 kDa periplasmic protein that is tethered to the cytoplasmic membrane by a single transmembranous helix [111]. Fully reduced Cox11 of *Sinorhizobium meliloti* is monomeric, but in the absence of reductants, forms a dimer which binds two copper ions at the dimer interface [112]. Cox11 is required for the assembly of the Cu_B center of aa_3-type, but not cbb_3-type COX [113]. In *P. denitrificans*, it was shown that the Cox11-homologue, CtaG, interacts directly with COX [110]. In mitochondria, Cox11 receives Cu from the small soluble copper chaperone Cox17, but it is currently unknown whether copper loading of Cox11 in bacteria requires additional proteins [114].

The second periplasmic protein that appears to universally participates in COX assembly in bacteria as well as in mitochondria is ScoI (also called PrrC, or Sco1 in eukaryotes). ScoI-type proteins can participate in the assembly of Cu_A as well as Cu_B sites and are dispensable under conditions of copper excess [110]. In *B. japonicum*, ScoI is a 25 kDa protein which possesses an N-terminal membrane anchor and is involved in the assembly of the Cu_A site on subunit II of the aa_3-type COX [109]. ScoI can bind Cu^+ or Cu^{2+} via two cysteine residues, but transfer of the copper to COX remains to be demonstrated [115]. In *R. capsulatus*, the Sco1 homologue SenC participates in cbb_3-type COX assembly under low copper conditions [116], and it was suggested that COX subunit I and SenC cooperate in the assembly of the Cu_B center of cbb_3-type COX [117]. For this process, SenC seems to require a PCu_AC-type (PcuC) copper chaperone which interacts with ScoI [118]. PCu_AC is a periplasmic protein that is present in many bacteria, but not in mitochondria. Similarly, in *R. capsulatus*, a PCu_AC-homologue, PccA, is required for Cu_B-center assembly of cbb_3-type COX and the maintenance of steady-state activity under low-copper conditions [119]. By chemical crosslinking, SenC and PccA were shown to form a complex during cbb_3-type COX assembly. However, the individual effects of SenC/ScoI and PccA/PCu_AC on cbb_3-type COX assembly are largely unexplored. PCu_AC adopts a cupredoxin-like fold, and probably coordinates Cu^+ via two methionine and two histidine residues. The formation of Cu_A centers also requires the function of the Sco1 and PCu_AC (hence its name) [120]. It was shown that PCu_AC of *Thermus thermophilus* metallates the Cu_A center of the ba_3-type COX in vitro, but the in vivo function remains unclear [121]. Finally, recent work in *Bradyrhizobium japonicum* led to the identification of a sulfide reductase, TlpA. It is a membrane–anchored thioredoxin that prepares Sco1 and cytochrome oxidase subunit II for copper insertion by reducing critical cysteine residues [122].

Clearly, the insertion of copper into Cu_A and Cu_B sites of bacterial COX enzymes is far from understood. There may be additional cofactors or chaperones involved. Also, it not yet clear how copper is specifically transferred from a dedicated CopA2-type copper exporting ATPase to the helper proteins for metallation of COX. That copper for insertion into COX is recruited from the cytoplasm might serve in the specificity of the process and prevent the mis-metallation of COX by Zn^{2+} or other heavy metals.

4.8.4 Copper Loading of Cuproenzymes in Cyanobacteria

4.8.4.1 Metallation of CucA

A nice example of copper loading of a protein in the periplasm is CucA of *Synechocystis* PCC 6803. CucA was identified as the most abundant copper protein in the cytoplasm of *Synechocystis* [123]. CucA is a cupin (a small β-barrel protein) that binds Cu^{2+} via a His_3-Glu_1 ligand set [124]. The protein is exported into the periplasmic space in a copper-free, unfolded state by the general secretory pathway and metallation takes place in the periplasmic space. Very little metallation of CucA takes place in the absence of either one of the two copper ATPases PacS or CtaA, and only about 60% metallation occurs in the absence of the CopZ-like copper chaperone Atx1. The absence of CtaA, PacS, and Atx1 did, however, not affect mRNA levels for CucA. Also, CucA could not acquire copper directly from the copper pool of the periplasmic space. So the copper acquired by CucA must be exported from the cytoplasm into the periplasmic space and is then guided to CucA by Atx1 or another periplasmic copper chaperone. Such a mechanism most likely serves to prevent mis-metallation of CucA by zinc or another metal.

4.8.4.2 Metallation of Plastocyanin

Plastocyanin is a copper-containing electron transfer protein localized in the thylakoid lumen of cyanobacteria. The biogenesis of thylakoid membranes is complex and is linked to the cytoplasmic membrane. Also, the thylakoid lumen might form a continuum with the periplasmic space [125, 126]. Figure 4.8 shows the current concept of thylakoid biogenesis and a hypothetical scheme of how the biosynthesis of plastocyanin could proceed. The concept is supported by the primordial cyanobacterium *Gloeobacter violaceus*, which in lieu of an internal thylakoid membrane system contains photosynthetically active patches in the cytoplasmic membrane [127]. This suggests spatially separated biogenic and photosynthetically active membrane microdomains, likely to represent the evolutionary starting point for the development of an internal thylakoid membrane system.

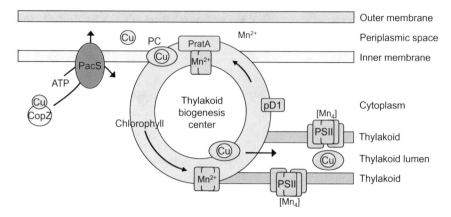

Fig. 4.8 Hypothetical scheme of the metallation of enyzmes in a thylakoid biogenesis center in *Synechocystis*. The photosystem II (PSII) assembly factor PratA accumulates in a membrane sub-fraction which also contains the precursor protein pD, other PSII precursor proteins and assembly factors, protochlorophyllide oxidoreductase, as well as the chlorophyll precursor chlorophyllide *a*, but not thylakoid marker proteins [127]. pD1 is loaded with periplasmatic manganese *via* PratA and delivers the Mn^{2+} to PSII, which is then further assembled and chlorophyll is inserted. Following further maturation steps, PSII traffics to the thylakoid membranes. Although plastocyanin (PC) is a soluble protein, it could similarly be assembled in the thylakoid biogenesis center as follows: CopZ delivers cytoplasmic Cu^+ to PacS for export to the periplasmic space, where it is specifically used for the metallation of *apo*-plastocyanin. Cu-plastocyanin then traffics to the thylakoid lumen, conceivably *via* a luminal compartment of the thylakoid biogenesis center that connects the periplasmic space to the thylakoid lumen. Additional assembly factors will be involved in the complex assembly process

4.8.5 Metallation of Methane Monooxygenase in Methanotrophs

Methanotrophic organisms require copper for the synthesis of membrane-bound methane monooxygenase (particulate, pMMO). The enzyme is a cuproenzyme which generally reduces methane to methanol, but other activities are known. pMMO is located in the cytoplasmic and/or in intracytoplasmic membranes, depending on the organism. Some methanotrophs can also express a soluble isoform of the enzyme (MMO) that is localized in the cytoplasm and uses iron as a cofactor. MMO is induced under copper-starvation to provide an alternative to the copper-requiring pMMO. The transition from iron-MMO to copper-pMMO appears to be controlled by a 'copper-switch', which involves various gene products such as MmoR, MmoG, MmoD, and methanobactin [128–130].

The metallation of pMMO with copper has not been investigated to a great extent and the description here is largely hypothetical. Conceivably, the intracellular membranes present in some methanogens form a continuum with the cytoplasmic membrane via a membrane biosynthesis center, similar to thylakoid membranes in cyanobacteria (see Sect. 4.8.4.2). Copper for the incorporation into pMMO is most likely taken up by methanobactin, which delivers copper to the cytoplasm (cf. Sect. 4.1.3). Cu^+ could then be transported by a dedicated copper ATPase or another transport system to the periplasm for delivery to *apo*-pMMO. This would be analogous to the mechanism that has previously been proposed for other organisms and for which there is some evidence. *M. trichosporium* OB3b, and probably most if not all methanogens, possess two putative copper ATPases, one of which could be dedicated to copper tolerance and the other to the metallation of cuproenzymes. But this mechanism of metallation of pMMO needs to be demonstrated.

4.9 An Fe–Cu Connection

In eukaryotic cells, homeostatic control mechanisms for iron are inextricably linked to Cu availability because they involve the action of multi-copper oxidases [131]. Recent findings suggests that a copper-iron connection might exist in bacteria as well. The thermophilic bacterium *Archaeoglobus fulgidus* uses iron in the intracellular Cu^+ distribution pathway. Sazinsky et al. [132] described an unusual CopZ copper chaperone (here called AfCopZ), very different from the previously described CopZ chaperones. AfCopZ features an N-terminal, typical CopZ-like Cu^+-binding domain, able to deliver Cu^+ to the copper ATPase CopA. However, AfCopZ also possesses an N terminal domain with a zinc-coordination site, followed by a [2Fe–2S] cluster that binds a single Cu atom and exhibits Cu^{2+} reductase activity. The AfCopZ N–terminus features a novel $\beta\alpha\alpha\beta\beta\beta\alpha$-fold that appears to be typical of extremophiles and is predicted to be produced as a separate polypeptide in organisms other than *A. fulgidus*. It was proposed that Cu^{2+} could bind transiently to an AfCopZ–site near the [2Fe–2S] cluster and be reduced to Cu^+. The cuprous ion could then be transferred intramolecularly to the N-terminal copper binding site for donation to a cognate copper homeostatic protein. Thus, *A. fulgidus* AfCopZ could provide coupling of Cu^{2+} reduction to Cu^+ chaperone function, facilitating Cu^+ transfer under extreme environmental conditions. Clearly, further work will be required to understand the function of this unusual copper chaperone.

4.10 Molybdenum Cofactor Synthesis and Copper

Molybdenum cofactor (Moco) is an essential prosthetic group for a range of pro- and eukaryotic oxidoreductases [133]. Moco is a molybdopterin that contains a molybdenum atom coordinated by a dithiolene group. In the two final steps of moco synthesis, molybdenum is added to the dithiolene sulphurs of molybdopterin via an adenylated intermediate, followed by conversion to moco. In *E. coli*, these last two steps are catalyzed by MogA and MoeA, in humans by Geph-1 and Geph-2, and in plants by the multifunctional enzyme Cnx1 (Cnx1G, Cnx1E). An X-ray structure of Cnx1G in complex with adenylated molybdopterin unexpectedly revealed a copper ion bound to the molybdopterin dithiolate sulphurs [134]. This novel moco-copper connection initially met with great excitement, as it would connect molybdenum and copper metabolisms and would provide the first evidence for a cytoplasmic copper-containing protein in bacteria. The copper bound to the dithiolene was considered as a possible leaving-group which would protect the sulfhydryls until they are occupied by the molybdenum. However, it was found that Cu^{2+}, Cd^{2+}, and AsO_2^{2-} could all bind nonspecifically to molybdopterin without the presence of either MogA or MoeA. Also, Cu^{2+} had a higher affinity for the dithiolene group of molybdopterin than molybdate and copper inhibited moco biosynthesis in *E. coli* [135]. Furthermore, it was found that the activity of the two moco-dependent enzymes dimethyl sulfoxide reductase and nitrate reductase of either *Escherichia coli* or *Rhodobacter sphaeroides* were not affected when copper was depleted from the media [136]. So the appearance of copper in molybdopterin biosynthesis may be fortuitous and serve no biological role.

References

1. Tottey S, Rich PR, Rondet SA et al (2001) Two Menkes-type atpases supply copper for photosynthesis in *Synechocystis* PCC 6803. J Biol Chem 276:19999–20004
2. Phung LT, Ajlani G, Haselkorn R (1994) P-type ATPase from the cyanobacterium *Synechococcus* 7942 related to the human Menkes and Wilson disease gene products. Proc Natl Acad Sci U S A 91:9651–9654
3. Lewinson O, Lee AT, Rees DC (2009) A P-type ATPase importer that discriminates between essential and toxic transition metals. Proc Natl Acad Sci U S A 106:4677–4682
4. Odermatt A, Krapf R, Solioz M (1994) Induction of the putative copper ATPases, CopA and CopB, of *Enterococcus hirae* by Ag^+ and Cu^{2+}, and Ag^+ extrusion by CopB. Biochem Biophys Res Commun 202:44–48
5. Koch HG, Winterstein C, Saribas AS et al (2000) Roles of the *ccoGHIS* gene products in the biogenesis of the *cbb$_3$*-type cytochrome *c* oxidase. J Mol Biol 297:49–65
6. Hassani BK, Astier C, Nitschke W et al (2010) CtpA a copper-translocating P-type ATPase involved in the biogenesis of multiple copper-requiring enzymes. J Biol Chem 285:19330–19337
7. Koh EI, Henderson JP (2015) Microbial copper-binding siderophores at the host-pathogen interface. J Biol Chem 290:18967–18974
8. Wilson BR, Bogdan AR, Miyazawa M et al (2016) Siderophores in iron metabolism: from mechanism to therapy potential. Trends Mol Med 22:1077–1090
9. Koh EI, Robinson AE, Bandara N et al (2017) Copper import in *Escherichia coli* by the yersiniabactin metallophore system. Nat Chem Biol 13:1016–1021
10. Noinaj N, Guillier M, Barnard TJ et al (2010) TonB-dependent transporters: regulation, structure, and function. Annu Rev Microbiol 64:43–60
11. Cavet JS (2014) Copper as a magic bullet for targeted microbial killing. Chem Biol 21:921–922
12. Solioz M (2016) Copper oxidation state and mycobacterial infection. Mycobact Dis 6:210–213
13. Kim HJ, Graham DW, DiSpirito AA et al (2004) Methanobactin, a copper-acquisition compound from methane-oxidizing bacteria. Science 305:1612–1615
14. Kenney GE, Rosenzweig AC (2013) Genome mining for methanobactins. BMC Biol 11:17
15. DiSpirito AA, Semrau JD, Murrell JC et al (2016) Methanobactin and the link between copper and bacterial methane oxidation. Microbiol Mol Biol Rev 80:387–409
16. Balasubramanian R, Kenney GE, Rosenzweig AC (2011) Dual pathways for copper uptake by methanotrophic bacteria. J Biol Chem 286:37313–37319

17. Ve T, Mathisen K, Helland R et al (2012) The *Methylococcus capsulatus* (Bath) secreted protein, MopE*, binds both reduced and oxidized copper. PLoS ONE 7:e43146

18. Zhang XX, Rainey PB (2008) Regulation of copper homeostasis in *Pseudomonas fluorescens* SBW25. Environ Microbiol 10:3284–3294

19. Wijekoon CJ, Young TR, Wedd AG et al (2015) CopC protein from *Pseudomonas fluorescens* SBW25 features a conserved novel high-affinity Cu(II) binding site. Inorg Chem 54:2950–2959

20. Lawton TJ, Kenney GE, Hurley JD et al (2016) The CopC family: structural and bioinformatic insights into a diverse group of periplasmic copper binding proteins. Biochemistry 55:2278–2290

21. Hirooka K, Edahiro T, Kimura K et al (2012) Direct and indirect regulation of the *ycnKJI* operon involved in copper uptake through two transcriptional repressors, YcnK and CsoR, in *Bacillus subtilis*. J Bacteriol 194:5675–5687

22. Cha JS, Cooksey DA (1993) Copper hypersensitivity and uptake in *Pseudomonas syringae* containing cloned components of the copper resistance operon. Appl Environ Microbiol 59:1671–1674

23. Gu W, Farhan Ul Haque M, Semrau JD (2017) Characterization of the role of *copCD* in copper uptake and the "copper-switch" in *Methylosinus trichosporium* OB3b. FEMS Microbiol Lett 164:fnx094

24. Kanamaru K, Kashiwagi S, Mizuno T (1994) A copper-transporting P-type ATPase found in the thylakoid membrane of the cyanobacterium *Synechococcus* species PCC7942. Mol Microbiol 13:369–377

25. Verméglio A, Lavergne J, Rappaport F (2016) Connectivity of the intracytoplasmic membrane of *Rhodobacter sphaeroides*: a functional approach. Photosynth Res 127:13–24

26. Niederman RA (2016) Development and dynamics of the photosynthetic apparatus in purple phototrophic bacteria. Biochim Biophys Acta 1857:232–246

27. Ekici S, Yang H, Koch HG et al (2012) Novel transporter required for biogenesis of *cbb*3-type cytochrome *c* oxidase in *Rhodobacter capsulata*. MBio 3:e00293-11

28. Ekici S, Turkarslan S, Pawlik G et al (2014) Intracytoplasmic copper homeostasis controls cytochrome *c* oxidase production. MBio 5

29. Wang Y, Hodgkinson V, Zhu S et al (2011) Advances in the understanding of mammalian copper transporters. Adv Nutr 2:129–137

30. Beaudoin J, Ioannoni R, Lopez-Maury L et al (2011) Mfc1 is a novel forespore membrane copper transporter in meiotic and sporulating cells. J Biol Chem 286:34356–34372

31. Beaudoin J, Ioannoni R, Labbe S (2012) Mfc1 is a novel copper transporter during meiosis. Commun Integr Biol 5:118–121

32. Beaudoin J, Ekici S, Daldal F et al (2013) Copper transport and regulation in *Schizosaccharomyces pombe*. Biochem Soc Trans 41:1679–1686

33. Khalfaoui-Hassani B, Verissimo AF, Koch HG et al (2016) Uncovering the transmembrane metal binding site of the novel bacterial major facilitator superfamily-type copper importer CcoA. MBio 7:e01981-15

34. Meydan S, Klepacki D, Karthikeyan S et al (2017) Programmed ribosomal frameshifting generates a copper transporter and a copper chaperone from the same gene. Mol Cell 65:207–219

35. Drees SL, Klinkert B, Helling S et al (2017) One gene, two proteins: coordinated production of a copper chaperone by differential transcript formation and translational frameshifting in *Escherichia coli*. Mol Microbiol 106:635–645

36. Gupta A, Lutsenko S (2009) Human copper transporters: mechanism, role in human diseases and therapeutic potential. Future Med Chem 1:1125–1142

37. Outten FW, Outten CE, Hale J et al (2000) Transcriptional activation of an *Escherichia coli* copper efflux regulon by the chromosomal MerR homologue, CueR. J Biol Chem 275:31024–31029

38. Djoko KY, Chong LX, Wedd AG et al (2010) Reaction mechanisms of the multicopper oxidase CueO from *Escherichia coli* support its functional role as a cuprous oxidase. J Am Chem Soc 132:2005–2015

39. Grass G, Thakali K, Klebba PE et al (2004) Linkage between catecholate siderophores and the multicopper oxidase CueO in *Escherichia coli*. J Bacteriol 186:5826–5833
40. Tree JJ, Kidd SP, Jennings MP et al (2005) Copper sensitivity of *cueO* mutants of *Escherichia coli* K-12 and the biochemical suppression of this phenotype. Biochem Biophys Res Commun 328:1205–1210
41. Grass G, Rensing C (2001) CueO is a multi-copper oxidase that confers copper tolerance in *Escherichia coli*. Biochem Biophys Res Commun 286:902–908
42. Singh SK, Grass G, Rensing C et al (2004) Cuprous oxidase activity of CueO from *Escherichia coli*. J Bacteriol 186:7815–7817
43. Wu T, Wang S, Wang Z et al (2015) A multicopper oxidase contributes to the copper tolerance of *Brucella melitensis* 16M. FEMS Microbiol Lett 362:1–7
44. Mancini S, Kumar R, Mishra V et al (2017) *Desulfovibrio* DA2_CueO is a novel multicopper oxidase with cuprous, ferrous, and phenol oxidase activity. Microbiol 163:1229–1236
45. Rowland JL, Niederweis M (2013) A multicopper oxidase is required for copper resistance in *Mycobacterium tuberculosis*. J Bacteriol 195:3724–3733
46. Wagner D, Maser J, Moric I et al (2006) Elemental analysis of the Mycobacterium avium phagosome in Balb/c mouse macrophages. Biochem Biophys Res Commun 344:1346–1351
47. Hodgkinson V, Petris MJ (2012) Copper homeostasis at the host-pathogen interface. J Biol Chem 287:13549–13555
48. Hasan NM, Lutsenko S (2012) Regulation of copper transporters in human cells. Curr Top Membr 69:137–161
49. Achard ME, Tree JJ, Holden JA et al (2010) The multi-copper oxidase CueO of *Salmonella enterica* serovar Typhimurium is required for systemic virulence. Infect Immun 78:2312–2319
50. Osman D, Cavet JS (2011) Metal sensing in *Salmonella*: implications for pathogenesis. Adv Microb Physiol 58:175–232
51. Huston WM, Jennings MP, McEwan AG (2002) The multicopper oxidase of *Pseudomonas aeruginosa* is a ferroxidase with a central role in iron acquisition. Mol Microbiol 45:1741–1750
52. Huston WM, Naylor J, Cianciotto NP et al (2008) Functional analysis of the multi-copper oxidase from *Legionella pneumophila*. Microbes Infect 10:497–503
53. Rodriguez-Montelongo L, Volentini SI, Farias RN et al (2006) The Cu(II)-reductase NADH dehydrogenase-2 of *Escherichia coli* improves the bacterial growth in extreme copper concentrations and increases the resistance to the damage caused by copper and hydroperoxide. Arch Biochem Biophys 451:1–7
54. Volentini SI, Farias RN, Rodriguez-Montelongo L et al (2011) Cu(II)-reduction by *Escherichia coli* cells is dependent on respiratory chain components. Biometals 24:827–835
55. Abicht HK, Gonskikh Y, Gerber SD et al (2013) Non-enzymatic copper reduction by menaquinone enhances copper toxicity in *Lactococcus lactis* IL1403. Microbiol 159:1190–1197
56. Mermod M, Magnani D, Solioz M et al (2012) The copper-inducible ComR (YcfQ) repressor regulates expression of ComC (YcfR), which affects copper permeability of the outer membrane of *Escherichia coli*. Biometals 25:33–43
57. Yu Z, Reichheld SE, Savchenko A et al (2010) A comprehensive analysis of structural and sequence conservation in the TetR family transcriptional regulators. J Mol Biol 400:847–864
58. Padilla-Benavides T, George Thompson AM, McEvoy MM et al (2014) Mechanism of ATPase-mediated Cu+ export and delivery to periplasmic chaperones: the interaction of *Escherichia coli* CopA and CusF. J Biol Chem 289:20492–20501
59. Kim EH, Nies DH, McEvoy MM et al (2011) Switch or funnel: how RND-type transport systems control periplasmic metal homeostasis. J Bacteriol 193:2381–2387
60. Long F, Su CC, Zimmermann MT et al (2010) Crystal structures of the CusA efflux pump suggest methionine-mediated metal transport. Nature 467:484–488
61. Lei HT, Bolla JR, Bishop NR et al (2014) Crystal structures of CusC review conformational changes accompanying folding and transmembrane channel formation. J Mol Biol 426:403–411

62. Su CC, Long F, Lei HT et al (2012) Charged amino acids (R83, E567, D617, E625, R669, and K678) of CusA are required for metal ion transport in the Cus efflux system. J Mol Biol 422:429–441
63. Chacon KN, Mealman TD, McEvoy MM et al (2014) Tracking metal ions through a Cu/Ag efflux pump assigns the functional roles of the periplasmic proteins. Proc Natl Acad Sci U S A 111:15373–15378
64. Patel SJ, Padilla-Benavides T, Collins JM et al (2014) Functional diversity of five homologous Cu$^+$-ATPases present in *Sinorhizobium meliloti*. Microbiol 160:1237–1251
65. Rademacher C, Masepohl B (2012) Copper-responsive gene regulation in bacteria. Microbiol 158:2451–2464
66. Giner-Lamia J, Lopez-Maury L, Reyes JC et al (2012) The CopRS two-component system is responsible for resistance to copper in the cyanobacterium *Synechocystis* sp. PCC 6803. Plant Physiol 159:1806–1818
67. Schelder S, Zaade D, Litsanov B et al (2011) The two-component signal transduction system CopRS of *Corynebacterium glutamicum* is required for adaptation to copper-excess stress. PLoS ONE 6:e22143
68. Hu YH, Wang HL, Zhang M et al (2009) Molecular analysis of the copper-responsive CopRSCD of a pathogenic *Pseudomonas fluorescens* strain. J Microbiol 47:277–286
69. Bansal-Mutalik R, Nikaido H (2014) Mycobacterial outer membrane is a lipid bilayer and the inner membrane is unusually rich in diacyl phosphatidylinositol dimannosides. Proc Natl Acad Sci U S A 111:4958–4963
70. Speer A, Rowland JL, Haeili M et al (2013) Porins increase copper susceptibility of *Mycobacterium tuberculosis*. J Bacteriol 195:5133–5140
71. Ward SK, Hoye EA, Talaat AM (2008) The global responses of *Mycobacterium tuberculosis* to physiological levels of copper. J Bacteriol 190:2939–2946
72. Wolschendorf F, Ackart D, Shrestha TB et al (2011) Copper resistance is essential for virulence of *Mycobacterium tuberculosis*. Proc Natl Acad Sci U S A 108:1621–1626
73. Darwin KH (2015) *Mycobacterium tuberculosis* and copper: a newly appreciated defense against an old foe? J Biol Chem 290:18962–18966
74. Fahey RC, Brown WC, Adams WB et al (1978) Occurrence of glutathione in bacteria. J Bacteriol 133:1126–1129
75. Newton GL, Arnold K, Price MS et al (1996) Distribution of thiols in microorganisms: mycothiol is a major thiol in most actinomycetes. J Bacteriol 178:1990–1995
76. Gaballa A, Newton GL, Antelmann H et al (2010) Biosynthesis and functions of bacillithiol, a major low-molecular-weight thiol in Bacilli. Proc Natl Acad Sci U S A 107:6482–6486
77. Kim EK, Cha CJ, Cho YJ et al (2008) Synthesis of γ-glutamylcysteine as a major low-molecular-weight thiol in lactic acid bacteria *Leuconostoc* spp. Biochem Biophys Res Commun 369:1047–1051
78. Schafer FQ, Buettner GR (2001) Redox environment of the cell as viewed through the redox state of the glutathione disulfide/glutathione couple. Free Radic Biol Med 30:1191–1212
79. Obeid MH, Oertel J, Solioz M et al (2016) Mechanism of attenuation of uranyl toxicity by glutathione in *Lactococcus lactis*. Appl Environ Microbiol 82:3563–3571
80. Helbig K, Bleuel C, Krauss GJ et al (2008) Glutathione and transition-metal homeostasis in *Escherichia coli*. J Bacteriol 190:5431–5438
81. Potter AJ, Trappetti C, Paton JC (2012) *Streptococcus pneumoniae* uses glutathione to defend against oxidative stress and metal ion toxicity. J Bacteriol 194:6248–6254
82. Latinwo LM, Donald C, Ikediobi C et al (1998) Effects of intracellular glutathione on sensitivity of *Escherichia coli* to mercury and arsenite. Biochem Biophys Res Commun 242:67–70
83. Vasak M, Meloni G (2011) Chemistry and biology of mammalian metallothioneins. J Biol Inorg Chem 16:1067–1078
84. Huckle JW, Morby AP, Turner JS et al (1993) Isolation of a prokaryotic metallothionein locus and analysis of transcriptional control by trace metal ions. Mol Microbiol 7:177–187
85. Gold B, Deng H, Bryk R et al (2008) Identification of a copper-binding metallothionein in pathogenic mycobacteria. Nat Chem Biol 4:609–616

86. Vita N, Platsaki S, Basle A et al (2015) A four-helix bundle stores copper for methane oxidation. Nature 525:140–143
87. Vita N, Landolfi G, Basle A et al (2016) Bacterial cytosolic proteins with a high capacity for Cu(I) that protect against copper toxicity. Sci Rep 6:39065
88. Baslé A, Platsaki S, Dennison C (2017) Visualizing copper storage: the importance of thiolate-coordinated tetranuclear clusters. Angew Chem Int Ed Engl 56:8697–8700
89. Rao NN, Gomez-Garcia MR, Kornberg A (2009) Inorganic polyphosphate: essential for growth and survival. Annu Rev Biochem 78:605–647
90. Hashemi F, Leppard GG, Kushner DJ (1994) Copper resistance in *Anabaena variabilis*: effects of phosphate nutrition and polyphosphate bodies. Microb Ecol 27:159–176
91. Keasling JD, Hupf GA (1996) Genetic manipulation of polyphosphate metabolism affects cadmium tolerance in *Escherichia coli*. Appl Microbiol Biotechnol 62:743–746
92. Keasling JD (1997) Regulation of intracellular toxic metals and other cations by hydrolysis of polyphosphate. Ann N Y Acad Sci 829:242–249
93. Van Veen HW, Abee T, Kortstee GJ et al (1994) Generation of a proton motive force by the excretion of metal-phosphate in the polyphosphate-accumulating *Acinetobacter johnsonii* strain 210A. J Biol Chem 269:29509–29514
94. Grillo-Puertas M, Schurig-Briccio LA, Rodriguez-Montelongo L et al (2014) Copper tolerance mediated by polyphosphate degradation and low-affinity inorganic phosphate transport system in *Escherichia coli*. BMC Microbiol 14:72
95. Alvarez S, Jerez CA (2004) Copper ions stimulate polyphosphate degradation and phosphate efflux in *Acidithiobacillus ferrooxidans*. Appl Environ Microbiol 70:5177–5182
96. Remonsellez F, Orell A, Jerez CA (2006) Copper tolerance of the thermoacidophilic archaeon *Sulfolobus metallicus*: possible role of polyphosphate metabolism. Microbiol 152:59–66
97. Scheuring S, Nevo R, Liu LN et al (2014) The architecture of *Rhodobacter sphaeroides* chromatophores. Biochim Biophys Acta 1837:1263–1270
98. Stolle P, Hou B, Brüser T (2016) The Tat substrate CueO Is transported in an incomplete folding state. J Biol Chem 291:13520–13528
99. Osman D, Patterson CJ, Bailey K et al (2013) The copper supply pathway to a *Salmonella* Cu, Zn-superoxide dismutase (SodCII) involves P1B-type ATPase copper-efflux and periplasmic CueP. Mol Microbiol 87:466–477
100. Pontel LB, Audero ME, Espariz M et al (2007) GolS controls the response to gold by the hierarchical induction of *Salmonella*-specific genes that include a CBA efflux-coding operon. Mol Microbiol 66:814–825
101. Checa SK, Espariz M, Audero ME et al (2007) Bacterial sensing of and resistance to gold salts. Mol Microbiol 63:1307–1318
102. Osman D, Waldron KJ, Denton H et al (2010) Copper homeostasis in *Salmonella* is atypical and copper-CueP is a major periplasmic metal complex. J Biol Chem 285:25259–25268
103. Yoon BY, Kim YH, Kim N et al (2013) Structure of the periplasmic copper-binding protein CueP from *Salmonella enterica* serovar Typhimurium. Acta Crystallogr D Biol Crystallogr 69:1867–1875
104. Brzezinski P, Gennis RB (2008) Cytochrome c oxidase: exciting progress and remaining mysteries. J Bioenerg Biomembr 40:521–531
105. Solioz M (2018) Copper and bacteria. Elsevier, Amsterdam
106. Raimunda D, Padilla-Benavides T, Vogt S et al (2013) Periplasmic response upon disruption of transmembrane Cu transport in *Pseudomonas aeruginosa*. Metallomics 5:144–151
107. Raimunda D, Gonzalez-Guerrero M, Leeber BW III et al (2011) The transport mechanism of bacterial Cu$^+$-ATPases: distinct efflux rates adapted to different function. Biometals 24:467–475
108. Gonzalez-Guerrero M, Raimunda D, Cheng X et al (2010) Distinct functional roles of homologous Cu$^+$ efflux ATPases in *Pseudomonas aeruginosa*. Mol Microbiol 78:1246–1258
109. Buhler D, Rossmann R, Landolt S et al (2010) Disparate pathways for the biogenesis of cytochrome oxidases in *Bradyrhizobium japonicum*. J Biol Chem 285:15704–15713

110. Gurumoorthy P, Ludwig B (2015) Deciphering protein-protein interactions during the biogenesis of cytochrome *c* oxidase from *Paracoccus denitrificans*. FEBS J 282:537–549
111. Carr HS, Maxfield AB, Horng YC et al (2005) Functional analysis of the domains in Cox11. J Biol Chem 280:22664–22669
112. Banci L, Bertini I, Cantini F et al (2004) Solution structure of Cox11, a novel type of beta-immunoglobulin-like fold involved in CuB site formation of cytochrome c oxidase. J Biol Chem 279:34833–34839
113. Thompson AK, Smith D, Gray J et al (2010) Mutagenic analysis of Cox11 of *Rhodobacter sphaeroides*: insights into the assembly of Cu(B) of cytochrome c oxidase. Biochemistry 49:5651–5661
114. Horng YC, Cobine PA, Maxfield AB et al (2004) Specific copper transfer from the Cox17 metallochaperone to both Sco1 and Cox11 in the assembly of yeast cytochrome C oxidase. J Biol Chem 279:35334–35340
115. Balatri E, Banci L, Bertini I et al (2003) Solution structure of Sco1: a thioredoxin-like protein Involved in cytochrome *c* oxidase assembly. Structure 11:1431–1443
116. Lohmeyer E, Schroder S, Pawlik G et al (2012) The ScoI homologue SenC is a copper binding protein that interacts directly with the *cbb*3-type cytochrome oxidase in *Rhodobacter capsulata*. Biochim Biophys Acta 1817:2005–2015
117. Thompson AK, Gray J, Liu A et al (2012) The roles of *Rhodobacter sphaeroides* copper chaperones PCu(A)C and Sco (PrrC) in the assembly of the copper centers of the aa(3)-type and the cbb(3)-type cytochrome c oxidases. Biochim Biophys Acta 1817:955–964
118. Serventi F, Youard ZA, Murset V et al (2012) Copper starvation-inducible protein for cytochrome oxidase biogenesis in *Bradyrhizobium japonicum*. J Biol Chem 287:38812–38823
119. Trasnea PI, Utz M, Khalfaoui-Hassani B et al (2016) Cooperation between two periplasmic copper chaperones is required for full activity of the *cbb* -type cytochrome *c* oxidase and copper homeostasis in *Rhodobacter capsulatus*. Mol Microbiol 100:345–361
120. Banci L, Bertini I, Ciofi-Baffoni S et al (2005) A copper(I) protein possibly involved in the assembly of CuA center of bacterial cytochrome c oxidase. Proc Natl Acad Sci U. S. A 102:3994–3999
121. Abriata LA, Banci L, Bertini I et al (2008) Mechanism of Cu(A) assembly. Nat Chem Biol 4:599–601
122. Abicht HK, Scharer MA, Quade N et al (2014) How periplasmic thioredoxin TlpA reduces bacterial copper chaperone ScoI and cytochrome oxidase subunit II (CoxB) prior to metallation. J Biol Chem 289:32431–32444
123. Tottey S, Waldron KJ, Firbank SJ et al (2008) Protein-folding location can regulate manganese-binding versus copper- or zinc-binding. Nature 455:1138–1142
124. Waldron KJ, Firbank SJ, Dainty SJ et al (2010) Structure and metal-loading of a soluble periplasm cupro-protein. J Biol Chem 285:32504–32511
125. Rast A, Heinz S, Nickelsen J (2015) Biogenesis of thylakoid membranes. Biochim Biophys Acta 1847:821–830
126. Frain KM, Gangl D, Jones A et al (2016) Protein translocation and thylakoid biogenesis in cyanobacteria. Biochim Biophys Acta 1857:266–273
127. Rexroth S, Mullineaux CW, Ellinger D et al (2011) The plasma membrane of the cyanobacterium *Gloeobacter violaceus* contains segregated bioenergetic domains. Plant Cell 23:2379–2390
128. Scanlan J, Dumont MG, Murrell JC (2009) Involvement of MmoR and MmoG in the transcriptional activation of soluble methane monooxygenase genes in *Methylosinus trichosporium* OB3b. FEMS Microbiol Lett 301:181–187
129. Semrau JD, Jagadevan S, DiSpirito AA et al (2013) Methanobactin and MmoD work in concert to act as the 'copper-switch' in methanotrophs. Environ Microbiol 15:377–386
130. Kenney GE, Sadek M, Rosenzweig AC (2016) Copper-responsive gene expression in the methanotroph *Methylosinus trichosporium* OB3b. Metallomics 8:931–940
131. Fox PL (2003) The copper-iron chronicles: the story of an intimate relationship. Biometals 16:9–40

132. Sazinsky MH, LeMoine B, Orofino M et al (2007) Characterization and structure of a Zn^{2+} and [2Fe-2S]-containing copper chaperone from *Archaeoglobus fulgidus*. J Biol Chem 282:25950–25959
133. Schwarz G, Mendel RR (2006) Molybdenum cofactor biosynthesis and molybdenum enzymes. Annu Rev Plant Biol 57:623–647
134. Kuper J, Llamas A, Hecht HJ et al (2004) Structure of the molybdopterin-bound Cnx1G domain links molybdenum and copper metabolism. Nature 430:803–806
135. Iobbi-Nivol C, Leimkuhler S (2013) Molybdenum enzymes, their maturation and molybdenum cofactor biosynthesis in *Escherichia coli*. Biochim Biophys Acta 1827:1086–1101
136. Morrison MS, Cobine PA, Hegg EL (2007) Probing the role of copper in the biosynthesis of the molybdenum cofactor in *Escherichia coli* and *Rhodobacter sphaeroides*. J Biol Inorg Chem 12:1129–1139

Chapter 5
Conclusions

Abstract In this final chapter, some concepts which might be helpful to the experimentalist are summarized. Also, a number of hypothesis are formulated. They were born out of generalizations based on experimental findings, but are not proven to be of general validity. They may however help to conceptualize bacterial copper homeostasis.

Keywords Free copper · Tris buffer · Copper reduction · Mimetics · Toxicity
Chalkophores · Polyphosphate · Chaperones · Copper loading

In way of conclusion of this book, a series of hypotheses and summary statements are formulated. The hypotheses are based on the author experience and 'feel' for the field, acquired by three decades of research in the copper field. The hypotheses and summary statements are briefly discussed below, but are also presented in the previous chapters at the appropriate places. Of course, some of the hypothesis will prove to be wrong, but some of the concepts put forth here address some open questions or raise new ones and could help to advance copper research.

(1) The free copper concentration is unknown in most experiments.

Only free copper is biologically active. In biochemical experiments, a large fraction of the copper will be bound to components of the experiment, such as bacteria, buffer substances or media components. In growth experiments, mM concentrations of Cu-salts are often added to complex media. Most of the added copper will be bound to media components and will not be biologically active. The concentration of free copper 'seen' by cells will be in the μM range. The copper buffering effect of complex media is also evident by the observation that bacteria are much more copper-sensitive in synthetic media, which contain less substances that complex copper. Since media components vary from lab to lab, it is impossible to compare bacterial copper sensitivity between laboratories. To determine the free copper concentration, a copper-selective electrode or a copper indicator would have to be used, but this is rarely done for technical reasons. Alternatively, a copper buffer can be employed, as elegantly demonstrated by Changela et al. [1].

© The Author(s) 2018   81
M. Solioz, *Copper and Bacteria*, SpringerBriefs in Biometals,
https://doi.org/10.1007/978-3-319-94439-5_5

(2) Tris buffer influences experiments with copper.

Tris-Cl is most widely used buffer in biochemical experiments. However, Tris strongly binds copper to form a $Tris_2Cu^{II}$-complex [2]. This neutral copper complex is more membrane permeable than uncomplexed Cu^{2+}. In addition, the complex-formation lowers the effective free copper concentration in the experiment. Preferable is the use of so-called 'Good-buffers' [3]. The Cl^- anion can also affect experimental outcomes since it stabilizes Cu^+ under aerobic conditions.

(3) Cu^{2+} is non-enzymatically reduced to Cu^+ by cytoplasmic membranes.

Copper in the presence of oxygen is the Cu^{2+}-form. It was demonstrated for Gram-positive *L. lactis* as well as Gram-negative *E. coli* that Cu^{2+} is reduced to Cu^+ in an apparently non-enzymatic reaction by quinones in the membrane [4, 5]. This process may be universal and may also take place at the mitochondrial inner membrane. Since Cu^+ is more toxic to bacteria than Cu^{2+}, a key role of the CueO multicopper oxidases might be the oxidation not only of Cu^+ exported into the periplasmic space by the copper ATPase, but also the oxidation of Cu^+ resulting from reduction at the cytoplasmic membrane.

(4) Cu^+ and Ag^+ are mimetics.

In line with this concept, it was shown that Cu^+ and Ag^+ are transported at the same rate and with the same affinity by the CopB copper ATPase of *E. hirae* [6]. It was also shown for many copper-responsive regulators that they also respond to Ag^+. If Cu^+ and Ag^+ do not behave similarly in their effects on a cuproenzyme, it is most likely due to toxicity effects of Ag^+ on unrelated components or other binding sites on a protein.

(5) The main copper toxicity mechanism is not oxidative stress, but Fe co-factor replacement.

The concept of oxidative stress as the major copper toxicity mechanism has been repeated so often in the literature that it has become a truism. Recent data suggests that the toxicity of copper to bacteria is due to co-factor displacement from iron-sulfur cluster proteins (see Sect. 2.2). This does not preclude intracellular ROS generation, since the displacement of iron from iron–sulfur clusters leads to increased cytoplas-mic Fe^{2+}, which can catalyze Fenton chemistry and lead to ROS. However, wild-type cells can normally cope with the resulting oxidative stress; hypersensitivity to copper due to oxidative stress can be observed under special conditions, such as in mutants deficient in superoxide dismutases or repair enzymes for oxidative DNA damage [7].

(6) Copper uptake by chalkophores is a general mechanism.

How bacteria are studied in the laboratory is very unnatural. The growth media are optimized for fast growth or other desired phenotypes. Such conditions may suppress inherent bacterial functions, such as chalkophore production or the polyphosphate metabolism. Since bacteria in natural environments usually grow in communities with other organisms, one species may take up copper via a chalkophore produced by another organism, as is known from iron acquisition with siderophores.

(7)   Phosphate metabolism plays a major role in copper tolerance.

The formation of polyphosphate in the cytoplasm, its hydrolysis and the export of copper-phosphate complexes may play a major role in copper tolerance (see Sect. 4.7.3). However, such mechanisms have received little attention.

(8)   All P1B-type Cu-ATPases pump both, Cu^+ and Ag^+, and only in the direction cytoplasm to extracytoplasmic space.

This hypothesis derives from the available structural and functional information on P1B-type ATPases.

(9)   Truncated N-termini of copper ATPases can serve as copper chaperones.

E. coli does not encode a CopZ-like chaperone. Instead, the first metal binding domain of CopA can be synthesized as a truncated protein by a mechanism of programmed ribosomal frameshifting (see Sect. 4.2.). This mechanism appears to be present also in other bacteria and even in eukaryotes. Structures which could lead to programmed ribosomal frameshifting were detected in the N-terminus of the human Wilson copper ATPase, ATP7B [8].

(10)   There is no direct need for cytoplasmic copper.

All cuproenzymes are localized in the cytoplasmic membrane or the periplasm and are metallated at or outside of the cytoplasmic membrane. The copper is routed through the cytoplasm and delivered to these enzymes by specialized copper ATPases or other copper exporters. To this end, most cells contain at least two copper ATPases, one for copper tolerance and one for copper delivery to the periplasm for the metallation of cuproenzymes. This insures specific metallation with copper and not another transition metal. Copper-loading of CucA in the periplasm of *Synechocystis* is a beautiful example of this principle [9]. The Moco-copper connection, which would suggest a need for cytoplasmic copper, is probably fortuitous and not required for cell function (see Sect. 4.10.).

(11)   Cuproenzymes of internal membranes systems are metallated at the cytoplasmic membrane.

It becomes increasingly clear that intracytoplasmic membranes of cyanobacteria and methanotrophs form a continuum with the cytoplasmic membrane, connected via a special membrane biogenesis center. Also, the lumen of internal membrane systems probably forms a continuum with the periplasmic space. Cuproenzymes which eventually localize to internal membranes are probably metallated at or near the cytoplasmic membrane, just like cuproenzymes of the periplasm or the cytoplasmic membrane (see Sect. 4.8.4).

References

1. Changela A, Chen K, Xue Y et al (2003) Molecular basis of metal-ion selectivity and zeptomolar sensitivity by CueR. Science 301:1383–1387
2. McPhail DB, Goodman BA (1984) Tris buffer—a case for caution in its use in copper-containing systems. Biochem J 221:559–560
3. Good NE, Winget GD, Winter W et al (1966) Hydrogen ion buffers for biological research. Biochemistry 5:467–477
4. Abicht HK, Gonskikh Y, Gerber SD et al (2013) Non-enzymatic copper reduction by menaquinone enhances copper toxicity in *Lactococcus lactis* IL1403. Microbiology 159:1190–1197
5. Volentini SI, Farias RN, Rodriguez-Montelongo L et al (2011) Cu(II)-reduction by *Escherichia coli* cells is dependent on respiratory chain components. Biometals 24:827–835
6. Solioz M, Odermatt A (1995) Copper and silver transport by CopB-ATPase in membrane vesicles of *Enterococcus hirae*. J Biol Chem 270:9217–9221
7. Kimura T, Nishioka H (1997) Intracellular generation of superoxide by copper sulphate in *Escherichia coli*. Mutat Res 389:237–242
8. Meydan S, Klepacki D, Karthikeyan S et al (2017) Programmed ribosomal frameshifting generates a copper transporter and a copper chaperone from the same gene. Mol Cell 65:207–219
9. Waldron KJ, Firbank SJ, Dainty SJ et al (2010) Structure and metal-loading of a soluble periplasm cupro-protein. J Biol Chem 285:32504–32511

Index